*Just-In-Time Math for Engineers*

# Just-In-Time Math for Engineers

*by*
**Archibald L. Fripp**
**Jon B. Fripp**
**Michael L. Fripp**

 Newnes

Amsterdam   Boston   Heidelberg   London   New York   Oxford
Paris   San Diego   San Francisco   Singapore   Sydney   Tokyo

Newnes is an imprint of Elsevier Science.

Copyright © 2003, Elsevier Science (USA).  All rights reserved.

No part of this publication may be reproduced, stored in a retrieval system, or transmitted in any form or by any means, electronic, mechanical, photocopying, recording, or otherwise, without the prior written permission of the publisher.

 Recognizing the importance of preserving what has been written, Elsevier Science prints its books on acid-free paper whenever possible.

**Library of Congress Cataloging-in-Publication Data**

Fripp, Archibald L.
    Just in time math for engineers / Archibald L. Fripp, Jon B. Fripp, Michael L. Fripp
        p. cm.
    ISBN: 0-7506-7535-7
    1. Mathematics. I. Fripp, Jon, 1967- II. Fripp, Michael, 1970- III. Title.

QA37.3.F75  2003
510—dc21

2003054132

**British Library Cataloging-in-Publication Data**
A catalogue record for this book is available from the British Library.

The publisher offers special discounts on bulk orders of this book.
For information, please contact:

Manager of Special Sales
Elsevier Science
200 Wheeler Road
Burlington, MA 01803
Tel: 781-313-4700
Fax: 781-313-4880

For information on all Newnes publications available, contact our World Wide Web home page at: http://www.newnespress.com

10 9 8 7 6 5 4 3 2 1

Printed in the United States of America

# Contents

*Foreword* ................................................................................................. *ix*
*Preface* .................................................................................................... *xi*
*Acknowledgments* ................................................................................ *xii*
*What's on the CD-ROM?* ..................................................................... *xiii*

***Chapter 1: Math – The Basics*** ........................................................... ***1***

    Numbers ................................................................................................. 2

    Some Laws ............................................................................................. 4

    More About Numbers ........................................................................... 6

    Logarithms ............................................................................................ 13

    Number Bases ...................................................................................... 15

***Chapter 2: Functions*** ........................................................................ ***19***

    Plotting Functions ................................................................................ 23

    Distance ................................................................................................. 24

    Slope ..................................................................................................... 26

    Functions and Coordinate Points ....................................................... 26

    Translation of Coordinates ................................................................. 31

    Other Functions ................................................................................... 32

    Log Functions ...................................................................................... 33

***Chapter 3: Algebra*** ............................................................................ ***37***

    First-degree Polynomials ..................................................................... 43

    Second-degree Polynomials ................................................................ 46

    Examples ............................................................................................... 54

# Contents

**Chapter 4: Matrices** ....................................................................... **59**

    More than Two Variables ......................................................... 74

    Some Matrix Arithmetic .......................................................... 76

**Chapter 5: Trigonometry** ............................................................ **83**

    Radians ................................................................................... 88

    Some Special Relationships ....................................................... 89

    The Distance Equation ............................................................ 92

    Adding Trigonometric Functions .............................................. 94

    Summing Interior Angles of Triangles and Quadrangles ........... 99

    The Laws of Sines and of Cosines .......................................... 101

    Periodicity ............................................................................. 103

    Phase Differences .................................................................. 107

    Effect of Combining Waves with Small Frequency Differences ........... 110

    Dissipation ............................................................................ 111

    Vectors ................................................................................. 113

    Examples ............................................................................... 113

**Chapter 6: Calculus** ................................................................. **125**

    Attacking Calculus ................................................................ 125

    Differential Calculus .............................................................. 126

    Derivatives of Compound Functions ....................................... 137

    The Importance of Zero Values of Derivatives ........................ 138

    Other Topics ......................................................................... 143

    Integral Calculus ................................................................... 148

    Example 1:  Root Mean Square ............................................. 156

    Example 2:  Fourier Analysis ................................................. 159

    Example 3:  Another RMS example ........................................ 165

    Example 4:  Automobile Acceleration .................................... 168

    Example 5:  Pressure ............................................................ 171

    Example 6:  Economics ......................................................... 173

    Example 7:  Semiconductor Photon Detector ......................... 174

## *Chapter 7: Probability and Statistics* ............................................. *177*

Probability ............................................................................... 177

Distributions ............................................................................ 196

Pascal's Triangle ...................................................................... 199

Moments .................................................................................. 202

Moments in Statistics ............................................................... 204

Probability Distributions in Real Life ........................................ 210

Correlation .............................................................................. 213

Curve Fitting ........................................................................... 217

How Do You Know if the Die Is Honest?— the $\chi^2$ Test ........ 219

Confidence Interval .................................................................. 220

Probability and Statistics Examples .......................................... 221

## *Chapter 8: Differential Equations* ............................................... *227*

Complete and Independent Solutions ...................................... 235

A Quick Review ....................................................................... 237

Spring-Mass System with Damping .......................................... 239

Multiple Roots to the Characteristic Equation ......................... 243

Forced Oscillations .................................................................. 244

Direct Integration .................................................................... 248

Boundary Value Problems ........................................................ 250

Series Solutions to Differential Equations ................................ 252

Examples ................................................................................. 254

The Navier-Stokes Equations ................................................... 261

## *Chapter 9: Vector Calculus* ....................................................... *265*

Solving Scalar Values .............................................................. 269

The Dot Product of Vectors ..................................................... 270

The Cross Product of Vectors ................................................... 272

Combining Vector Products ...................................................... 275

Rules of Determinants ............................................................. 277

Vector Calculus ....................................................................... 280

The Gradient ........................................................................... 282

# Contents

Divergence and Curl ................................................................ 284

The Continuity Equation ........................................................ 287

Line Integrals ........................................................................ 288

The Curl and Back to the Line Integral ................................ 290

Orthogonal Curvilinear Coordinates...................................... 292

In-depth Example: Electromagnetic Theory .......................... 297

Gauss's Law .......................................................................... 301

The Lorentz Force ................................................................ 301

Force on a Current Loop ...................................................... 302

**Chapter 10: Computer Mathematics** ............................. 311

Computerized Data Analysis ................................................ 311

Finite Difference Technique .................................................. 312

Optimization Techniques ...................................................... 316

Finite Element Techniques .................................................... 318

Example: Experimental Data and Theoretical Equations ...... 323

**Chapter 11: Chaos** .......................................................... 325

**Appendix A: Some Useful Mathematical Tables** ........... 335

**About the Authors** .......................................................... 341

**Index** ............................................................................... 343

# *Foreword*

My definition of engineering is the application of physics and other branches of science to the creation of products and services that (hopefully) make the world a better place. In order to do this, an engineer must master the use of certain tools. Some of these tools are physical in nature, like the computer, but for the most part an engineer's toolkit consists of mental skills developed through study of math and science. Mathematics is at the core of engineering, and skill at math is one of the main determining factors in how far an engineer advances in his or her career.

However, although they might not like to admit it, many practicing engineers have forgotten—or are uncomfortable with—much of the mathematics they learned in school. That's where *Just-in-Time Math for Engineers* comes in. The "just-in-time" concept of inventory management is familiar to most engineers, and I think it's a good title. The book provides a quick math review or refresher just when you need it most. If you're changing jobs, tackling a new problem, or taking a course that requires dusting off your math skills, this book can help. The authors, all engineers from various fields, have done a good job of distilling the fundamentals and explaining the concepts clearly and succinctly, from an engineering point of view.

A word of advice: during my engineering career, I've watched the computer become an indispensable and ubiquitous desktop engineering tool. It's changed the very shape and nature of engineering in many cases. However, in my opinion, it's far too easy nowadays for engineers to "let the computer handle the math." When modeling or simulating an engineering problem, no matter what the field, you should always be sure you understand the fundamental, underlying mathematics, so you can do a reality check on the results. Every engineer needs a "just-in-time" math skills update every now and then, so keep this book handy on your shelf.

Jack W. Lewis, P.E.
Author of *Modeling Engineering Systems: PC-Based Techniques and Design Tools*
Newnes Press

# *Preface*

*Mathematics is the gate and the key to the sciences.* — Roger Bacon, 1276

This is the stuff we use. The subject matter in this book is what the authors use in their professional lives – controlling stream bank erosion across America, designing equipment for the bottom of oil wells, and probing the phenomena of microgravity to understand crystal growth. Surprisingly, the civil engineer working on flood control, the researcher probing the Earth for its bounty, and the scientist conducting experiments on the Space Shuttle use the same equations. Whether on the ground, under the ground, or in space, mathematics is universal.

Math books tend to be written with the intent to either impress colleagues or to offer step-by-step instructions like a cookbook. As a result, many math books get lost in a sea of equations, and the reader misses the big picture of the way concepts relate to each other and are applied to reality. We provide the basic understanding for the application of mathematical concepts. This is the book we wish we had when we started our engineering studies. We also intend this book to be easy reading for people outside the technical sciences. We hope you use this book as a stepping-stone for understanding our physical world.

Our primary audience is the working engineer who wants to review the tools of the profession. This book will also be valuable to the engineering technician trying to advance in the work arena, the MBA with a non-technical background working with technical colleagues, and the college student seeking a broader view of the mathematical world. We believe our approach—concepts without mathematical jargon—will also find an audience among non-technical people who want to understand their scientific and engineering friends.

Mathematics is not just an intellectually stimulating, esoteric subject. It is incredibly useful, as well as fun. We hope this book addresses the usefulness of math and, in doing so, provides intellectual stimulation.

# Acknowledgments

We start this section thanking Newton, Leibniz, Bernoulli, and all those brilliant folks who laid out this subject for us. We humbly try to follow in their footsteps.

We also acknowledge, more personally, those who had direct impact on our engineering careers. We recognize our math teachers—at home, in the classroom and in the workplace. Jon and Michael would like to thank their father, Archie, for sharing with them his love for mathematics and for its utility. Special thanks also go to those public school teachers, college professors, and colleagues who offered extra support in the early years of our careers: Joe Ritzenthaler, Donna Perger, Carolyn Gaertner, Jackie DeYoung, George Hagedorn, Jan Boal, Ron Copeland, and Joe Bisognano. We appreciate as well the work experience that not only forced us to learn more but also gave meaning to what we learned. The experience of making a calculation, performing an experiment, then seeing the measurements of that experiment match the initial calculation is exciting.

Specifically, we wish to thank those who helped make this book possible. We write, and we know what we mean, but can anyone else understand what we write? We gratefully acknowledge the editorial help of Jean Fripp, Daniel Fripp, and Deborah Fripp who spent many hours deciphering our writing. One of the authors is responsible for making most of the figures; however, the sketches are the work of Valeska Fripp, and we appreciate her help. Finally, we appreciate the help, encouragement, and, when needed, threats of our editor, Carol Lewis of Elsevier.

# What's on the CD-ROM?

Included on the accompanying CD-ROM:

- a fully searchable eBook version of the text in Adobe pdf form
- additional solved problems for each chapter
- in the "Extras" folder, several useful calculators and conversion tools

Refer to the Read-Me file on the CD-ROM for more detailed information on these files and their applications.

CHAPTER **1**

# *Math – The Basics*

*No knowledge can be certain if it is not based upon mathematics or upon some other knowledge which is itself based upon the mathematical sciences.* — **Leonardo da Vinci**

If you want to skip this chapter, do so. But it may be a while since you've thought about this stuff, and we hope that you'll at least skim through it. Consider the easy math as a final check on your hiking boots before you start climbing the more exotic trails. Tight bootlaces will keep you from falling on the slippery slopes, and a good mathematical foundation will do the same for your mathematical education. So, read on, my friend: this might even be interesting.

What is math? Perhaps, more to your interest, what is engineering math? Math is a thought process; it isn't something you find. You do not synthesize it in the laboratory or discover it emanating from space or hidden in a coal mine[1]. You will discover math in your mind. Is math more than a consistent set of operations that help us describe what is real, or is it an immutable truth? We'll let you decide.

Math is a tool created by us human creatures. We have made the rules and the rules work. The precise rules evolve with time. Numbers were used for thousands of years before zero became a mathematical entity. Math is also a language. It's the language that scientists and engineers use to describe nature and tell each other how to build bridges and land on Mars. (How the Romans ever made the 50,000-seat Roman Coliseum using just Roman numerals for math, we'll never know.) Of course, the field of mathematics is an expanding field. Study on, and you may add to this expanding field of knowledge.

It is key to remember that math isn't something you have to understand, because there is nothing to understand. Math is something you simply have to know how to use and to become comfortable using. Math is not poetry, where there is meaning hidden between the lines. Math is not art, which has purpose even if it is not applied. From an engineering point of view, math is just a tool. (Although some purists might disagree.) This book hopes to help you use this tool of math better.

Now, let's have a quick review of the basics.

---

[1]  As you may know, the element helium was first discovered in the sun and later found trapped in pockets in coal mines and oil deposits.

# Numbers

When we think of numbers we tend to think of integers such as 1, 4, or 5,280, all of which represent something tangible whether it's money, grades, or distance. The next thought would be negative numbers: –2, –44, or –382. Negative numbers represent the lack of something, such as my bank account near payday. And, of course, half-way between 3 and –3 is zero. Zero is special. Used as a place holder in a number such as 304, it signifies that there are no tens in this number[2], but as you'll see (and probably already know) zero has special properties when we start to use it in mathematical operations.

So far we've only mentioned integers (we call zero an integer). Before we can logically talk about other types of numbers, we need to define some basic mathematical operations.

*Equality*: The equals, =, sign means that the expression on its righthand side has the exact same value as the expression on its lefthand side.

Examples:     3 = 3
              5 = 5

*Addition*: It's what you do when you put two or more sets of numbers together. The combined number is called the sum. But please note, you can only add like things— that is, they must have the same unit of measurement. You cannot add your apples to the money in your account unless you sell them and convert apples to money before the addition. However, you can add apples to oranges, but the unit of the sum becomes fruit.

Example:     2 + 3 = 5

*Subtraction*: Some folks say that subtraction doesn't exist. They say it's just negative addition. It does exist in the minds of engineers and scientists, however, so we'll talk about it. Subtraction is what happens when I write checks on my bank account: the sum of money in the account decreases. If it's close to payday, and I keep writing checks when I have zero or fewer dollars in the account, the math still works. The bank balance just becomes more negative. If I should get very careless and write a check for a negative amount of money, the bank may not know how to handle it, but the mathematician just says that I'm trying to subtract a negative number, which is the same as adding that number. That is

$-(-2) = 2$ .

---

[2] Assuming that we're using base 10 arithmetic—you'll see more on this later in the chapter.

This is silly when applied to a bank account, but math is just a set of tools. The mathematical convention works.

Example:   $5 - 2 = 3$

*Multiplication*: Multiplication is just adding a bunch of times. When we get our final answer we call it the product. If six people put $5 each into my bank account, how much more money would I have? We could add $5 + 5 + 5 + 5 + 5 + 5$ to the account, or we could multiply $6 \times 5$. It's the same thing[3].

Units are still important in multiplication, but you have more flexibility. With that flexibility comes power, and with power comes danger. In addition, you must keep the same units on each item in the list that you are adding. You add apples to apples. If you add apples to oranges, you change the units to fruit. In multiplication, you multiply the units as well as the numbers. In our simple example, we multiplied six people times $5 per person. People times dollars/person just winds up as dollars. We'll talk a lot more about units as we go along.

*Division*: Division is the inverse of multiplication. Say we have 60 apples and ask how many apples we can give to four different people. We can count the apples out one by one to make four piles of fifteen apples each, or we can divide four into sixty In this example, 60 is the dividend, four is the divisor, and the result, fifteen, is the quotient.

And don't forget units. The units divide just as the numbers divided. We had 60 apples divided by four sets, so we get 15 apples per set. Perhaps a clearer example would be to determine the average speed of a car if a trip of 1200 kilometers required 15 hours driving time. The dividend is 1200 kilometers, the divisor is 15 hours, and the quotient is 80 kilometers per hour.

We can write this division problem as

$$15\overline{)1200}^{\,80} \quad \text{or} \quad 1200\,\text{km} \div 15\,\text{hrs} = 80\,\text{km/hr}$$

or we can write it as

$$\frac{1200\,\text{km}}{15\,\text{hrs}} = 80\,\text{km/hr} .$$

In this form, the number in the 1200 position is called the numerator, the number in the 15 position is the denominator, and the result is still the quotient.

---

[3] We're sure that you know this, but here goes anyway. The symbols for multiplication are $\times$, $\cdot$, *, or nothing. That is, the quantity $a$ multiplied by the quantity $b$ can be written as $a \times b$, $a * b$, $a \cdot b$, or $ab$. If we're multiplying a couple of numbers by another number, we might put the pair in parenthesis, like this: $ab + ac = a(b + c)$. We'll save trees and use nothing unless a symbol is needed for clarity.

If we should state a division problem where the numerator was a lower magnitude than the denominator, we call that expression a fraction. Of course, $\frac{3}{2}$, 3/2, and $\frac{1200}{15}$ are fractions, but we tend to think of 3/2 as 1 plus the fraction 1/2. We could also call any division problem a fraction. A fraction is the ratio between two numbers. It's just a convenient term to apply to a division problem.

The expression 6/8 is a fraction. You will doubtless recognize that 6/8 is the same as 3/4. We generally prefer to write this fraction as 3/4, which is expressed in its *lowest terms*.

And while on the subject of fractions, let's talk about the decimal. We can leave the fraction, 3/4 , as is, or recognize that it is equal to 75/100. This allows us to write it easily as 0.75, the decimal equivalent to the fraction. All you are doing when you convert from a fraction to a decimal is continue to divide even when the quotient is less than one. For example,

$$\frac{3}{4} = 4\overline{)3.00}^{\,0.75}$$

## Some Laws

When it comes to math, laws are the rule, and we must carefully follow them. Hopefully, by the time you finish this book, your understanding of engineering mathematics will be such that you innately do the right thing, and you will not feel encumbered by a rote set of rules.

Most of these laws will seem like common sense to you. We'll use symbols instead of numbers in discussing these laws. These symbols[4], *a, b, c,...,* can represent any number unless otherwise stated. It is not important to remember the names of these laws, but it is important to know the concepts.

### *Associative Law of Addition*

$$(a + b) + c = a + (b + c) \qquad\qquad \text{Eq. 1-1}$$

We use the parentheses to dictate the order in which the operations are performed. The operations within the parentheses are performed first. The Associative Law of Addition simply states that it doesn't matter which numbers you add first; the answer will be the same.

Example:

$(3 + 2) + 4 = 3 + (2 + 4)$ because

$(5) + 4 = 3 + (6)$.

---

[4]  When using symbols in lieu of numbers, we're doing algebra.

Note that the use of the parenthesis is redundant in the second step. We used it only to show the original grouping.

### *Associative Law of Multiplication*

$$(ab)c + a(bc) \hspace{6cm} \text{Eq. 1-2}$$

The Associative Law of Multiplication, like the law for addition, simply states that it doesn't matter which numbers you multiply first because the answer will be the same.

Example:

$$(2 \times 4) \times 5 = 2 \times (4 \times 5) \text{ because}$$

$$(8) \times 5 = 2 \times (20)$$

Again, the parentheses are redundant in the second step. We won't keep disclaiming this obvious fact as we step through the rules.

### *Commutative Law of Addition*

$$a + b = b + a \hspace{6cm} \text{Eq. 1-3}$$

Example:

$$(3 + 4) = (4 + 3) \text{ because (this is too easy)}$$

$$7 = 7$$

The Commutative Law of Addition states that order doesn't matter. You will get the same answer if you add the numbers forwards or if you add them backwards.

### *Commutative Law of Multiplication*

$$ab = ba \hspace{6cm} \text{Eq. 1-4}$$

Example:

$$3 \times 4 = 4 \times 3 \text{ because}$$

$$12 = 12$$

The Commutative Law of Multiplication, like the law for addition, states that order doesn't matter in multiplication either. Switching the order of the numbers will get you to the same answer in the end.

### *Distributive Law*

$$a(b + c) = ab + ac \hspace{6cm} \text{Eq. 1-5}$$

Example:

$$2(3 + 5) = 2 \times 3 + 2 \times 5 \text{ because}$$

$$2 \times (8) = 6 + 10$$

The Distributive Law just distributes the *a* between the *b* and the *c*. The parentheses mean that both the *b* and the *c* are multiplied by the *a*.

### *Rules Using Zero*

1)  $a + 0 = a$

    Example:

    $$7 + 0 = 7$$

2)  $a \times 0 = 0$

    Example:

    $$5 \times 0 = 0$$

3)  a/0 is undefined[5]

    Example:

    There is no example for this one.

## More About Numbers

### *Rational Numbers*

A rational number is any number that can be written in the form *a/b* as long as *b* is not zero. All integers are rational numbers, but so are 2/3, –3/22, and 51/2.

### *Irrational Numbers*

Don't be confused by the dictionary when it says that "rational" is based on reason and "irrational" is not. An irrational number is simply a number that can't be expressed as a fraction. Both rational and irrational numbers are real numbers—that is, they both correspond to a real point on a scale. Here's an example of an irrational number: try to find a number that when multiplied by itself equals two. It doesn't exist. Start with 1.414, multiply it by itself, and see that it's close[6] to two. Keep adding small amounts to your starting number, 1.414, and you'll get very close to two. You may get larger than two, but you'll never equal two.

---

[5]  Some might say a/0 equals infinity (if $a \neq 0$). That logic follows from the fact that $\frac{a}{x}$ becomes larger and larger in magnitude as *x* becomes smaller in magnitude (we say "magnitude" to not confuse the relative values of small positive numbers with large negative numbers). But if $\frac{a}{0}$ equals infinity, then does $\frac{2a}{0}$ equal two infinities? And should it be positive infinity or negative infinity? Too messy. Just stay away from a/(you know what).

[6]  Be wary of calculators that round off too quickly.

*Imaginary Numbers*

If you like irrational numbers, you'll love imaginary numbers. The big difference is that you can go through the rest of your engineering life[7] without worrying about the definition of irrational numbers (you'll just use them), but the definition of imaginary numbers will follow you forever.

An imaginary number doesn't exist, but engineers (and others) use it extensively to describe real things. This concept is better left to numerous examples sprinkled about in the more advanced sections of this book. For now, we'll just give an example and go on to an even more obtuse topic.

An example of an imaginary number is that number which, when multiplied by itself, is equal to –1. And, of course, such a number doesn't exist. You just learned that a negative number times any other negative number is a positive number—that's why it's an imaginary number.

The number we're talking about is called *i*. That is,

$$i \times i = -1.$$                    **Remember this one!**  Eq. 1-6

Likewise

$$2i \times 2i = -4.$$

To add even more intrigue to the world of imaginary numbers, electrical engineers use the letter "j" instead of "i" since they have reserved "i" for electrical current. So

$$\sqrt{-1} = i = j$$

*Complex Numbers*

A complex number is a combination of real and imaginary numbers. Example: The number 2 is a real number. The number $3i$, or $i3$, is an imaginary number. The number $2 + i3$ is a complex number.

While on this subject, let's look at one more item, the complex conjugate. If $2 + i3$ is our complex number, then $2 - i3$ is its complex conjugate. "Who cares?" We do. The complex conjugate can turn complex numbers into real numbers. Note what happens when we multiply a complex number by its complex conjugate,

$$(2+i3) \times (2-i3) = 2 \times 2 - 2 \times i3 + i3 \times 2 - i3 \times i3.$$

The terms with imaginary numbers cancel out, the $i3$ multiplied by itself becomes –9, and we are left with

$$(2+i3) \times (2-i3) = 2 \times 2 - (-9)$$
$$= 4 + 9 = 13$$

Hence, in symbols,

$$(a+ib)(a-ib) = a^2 + b^2$$        **Remember this one!**  Eq. 1-7

---

[7] Computer science may be an exception.

## *More Rules*

These rules are rather self evident, but we'll throw in a few examples. (In no case below do we, or should you, divide by zero.)

$a/_1 = a$ , that is, $3/_1 = 3$

$a/_b = c/_d$ if, and only if, $ad = bc$ . Example: $2/_7 = 8/_{28}$ , because $7 \cdot 8 = 2 \cdot 28$

$a/_c + b/_c = \dfrac{(a+b)}{c}$ . Example: $16/_4 + 20/_4 = \dfrac{(16+20)}{4}$ , because $4 + 5 = \dfrac{(36)}{4}$

$a/_b \cdot c/_d = \dfrac{ac}{bd}$ . Example: $20/_4 \cdot 30/_6 = \dfrac{20 \cdot 30}{4 \cdot 6}$ , because, $5 \cdot 5 = \dfrac{600}{24}$

$(a/_b) \div (c/_d) = \dfrac{ad}{bc}$ . Example: $40/_2 \div 8/_4 = \dfrac{40 \cdot 4}{2 \cdot 8}$ , because, $20 \div 2 = \dfrac{160}{16}$

If $a = b + c$ , then $a + d = b + c + d$ . Example: If $5 = 3 + 2$ , then $5 + 7 = 3 + 2 + 7$

If $a = bc$ , then $ad = bcd$ ; example: If $12 = 3 \cdot 4$ , then $12 \cdot 2 = (3 \cdot 4) \cdot 2$

The Cancellation Rule, $\dfrac{a}{c} \cdot \dfrac{c}{d} = a/_d$ . Example: $\dfrac{12}{4} \cdot \dfrac{4}{2} = \dfrac{12}{2}$

Are you bored yet? If not, here are some definitions.

## *Definitions*:

$1/_a$ is called the *reciprocal* of a. Example: $1/_2$ is the reciprocal of 2; likewise, 2 is the reciprocal of $1/_2$ . We sometimes refer to the reciprocal of a value as its inverse[8] value.

Equality, If $a = c$ , and $b = c$ , then $a = b$ .

Example: If $3 + 4 = 7$ , and $6 + 1 = 7$ , then $3 + 4 = 6 + 1$

## *Inequality signs* $(<, >, \leq, \geq)$

If $a$ and $b$ are both real numbers and if $a - b$ is a positive number, then $a$ is larger than $b$ (even if both numbers are negative). The relationship between $a$ and $b$ can be written as

$a > b$, or $b < a$

---

[8] In fact we'll use the term "inverse" in a variety of ways as we go along. It'll always be analogous to the use of the word here in that we'll take something and turn it upside down.

which is spoken as "*a* is greater than *b*" or "*b* is less than *a*." Example: $5 > 3$, or $3 < 5$.

If it is only known that *a* is not smaller than *b* (that is, *a* may equal *b* or be larger than *b*) then we write that relationship as

$$a \geq b$$

and say, "a is equal to or greater than *b*" or "*b* is less than or equal to *a*." Example: if $5 \geq a$, then *a* is either equal to 5 or it is less than 5.

### Some Inequality Relationships

The notation

$$a > b > c$$

means that *b* is less than *a* but greater than *c*. Example: $7 > 5 > 3$.

For any two numbers, only one of the following is true:

$a < b$, or $a = b$, or $a > b$. Example: $a < 5$, or $a = 5$, or $a > 5$.

If $a < b$ and $b < c$ then $a < c$. Example 1: Since $5 < 7$ and $7 < 9$ then $5 < 9$. Example 2: Since $-7 < -5$ and $-9 < -7$ then $-9 < -5$. (This gets silly.)

If $a < b$, then $a + c < b + c$ for any value of *c*. Example 1: Since $1 < 2$, then $1 + 235 < 2 + 235$. Example 2: Since $-2 < -1$, then $-2 + 235 < -1 + 235$.

If $c > 0$ and $a > b$, then $ac < bc$, regardless of the signs of *a* and *b*. Example 1: Since $2 > 0$, and $4 > 3$, then $2 \cdot 4 > 2 \cdot 3$. Example 2: Since $2 > 0$, and $-3 > -4$, then $2 \times (-3) > 2 \times (-4)$.

If $c < 0$ and $a < b$, then $ac > bc$. This is just the converse to the last one.

With a little bit of scratching about, you can show the same relationships as above for the "equal to or greater than" signs.

### Absolute Value

The absolute value of a number is its distance from zero, the origin of our counting scale. Sometimes the sign of the number is all important. If we're talking about my bank account, the difference between happiness and misery is whether or not I have plus one dollar or minus one dollar in the bank the day before payday. In other circumstances, the magnitude, or absolute value, is all important. Electrical engineers are most concerned with the magnitude of the voltage that they're handling and not just the sign in a relative sense. A 6-volt battery may have a "larger" value than a –20,000-volt cathode in a television set, but we know which one we'd least rather touch, and the emergency room physician will not be concerned about the voltage's sign when treating your burns.

The absolute value of $a$, or $|a|$, is defined as

$|a| = a$ *if* $a \geq 0$. Example: $|5| = 5$

and

$|a| = -a$ *if* $a \leq 0$. Example: $|-5| = -(-5) = 5$

Glance at the following facts about absolute values and convince yourself that they're correct.

$|-a| = |a|$. Example: $|-5| = -(-5) = 5$ and $|5| = 5$.

$|a| \geq 0$. Look at the last few examples.

$|0| = 0$.

$|ab| = |a| \times |b|$.

Example: $|-3 \times 4| = |-12| = |12| = 12$, and $|-3| \times |4| = |3| \times |4| = 3 \times 4 = 12$

$\left| \dfrac{a}{b} \right| = \dfrac{|a|}{|b|}$.

Example: $\left| \dfrac{-8}{2} \right| = |-4| = 4$, and $\dfrac{|-8|}{|2|} = \dfrac{|8|}{|2|} = \dfrac{8}{2} = 4$

and

$|a| * |a| = |a \times a| = a * a$. (We think this truth is too self-evident to warrant an example.)

Those facts about the absolute value were easy enough. Now scratch your head over the next two sets of inequalities and convince yourself that they are correct.

$|x| < c$ and $-c < x < c$.

Example: If $x = -5$, and $c = 7$, then $|-5| = 5 < 7$ and $-7 < -5 < 7$ are the exact same mathematical statements, as are

$|x - a| < c$ and $(a - c) < x < (a + c)$.

Example: If $x = 7$, $a = 5$, and $c = 3$, then $|7 - 5| = 2 < 3$, and $(5 - 3) < 7 < (5 + 3)$.

Work them out symbolically, if you can, and then verify your findings with numbers. Use both positive and negative numbers to show the universality of the relationships.

*Exponents*

You, of course, know how to multiply $a$ times $b$, which we write as $ab$. When multiplying $b$ times $b$, you can write it as $bb$ or $b^2$. We call the form $b^2$ "b to the second power" and we use the exponent form to write it.

The word "exponent" may sound a bit esoteric, but it's just a shorthand way to multiply and divide. We roughly described a simple exponent above. A more formal definition is

$$b^n = b \cdot b \cdot b \cdot b \dots \text{n times,}$$
**Remember this one!** Eq. 1-8

where $b$ is any number and $n$, the exponent, is a positive integer.

But this is not the complete definition because the exponent can be a negative number or even a fraction. However, staying for the moment with exponents as positive integers, you should be able to verify the following rules:

$$b^n b^m = b^{(n+m)}. \text{ Example: } 2^2 \cdot 2^3 = 4 \cdot 8 = 32 = 2^5$$

$$\left(b^n\right)^m = b^{nm}. \text{ Example: } \left(2^2\right)^3 = 4^3 = 64 = 2^{2 \cdot 3} = 2^6$$

$$(bc)^n = b^n c^n. \text{ Example: } (3 \cdot 4)^2 = 12^2 = 144 \text{ and } 3^2 \cdot 4^2 = 9 \cdot 16 = 144.$$

$$\left(\frac{b}{c}\right)^n = \frac{b^n}{c^n}. \text{ Example: } \left(\frac{8}{2}\right)^3 = 4^3 = 64, \text{ and } \frac{8^3}{2^3} = \frac{512}{8} = 64.$$

and

$$\frac{b^n}{b^m} = b^{n-m}. \text{ Example: } \frac{3^5}{3^2} = \frac{243}{9} = 27, \text{ and } 3^{(5-2)} = 3^3 = 27.$$

$$\frac{b^n}{b^m} = \frac{1}{b^{m-n}}. \text{ Example: } \frac{2^3}{2^5} = \frac{8}{32} = \frac{1}{4}, \text{ and } \frac{1}{2^{(5-3)}} = \frac{1}{2^2} = \frac{1}{4}.$$

or

$$\frac{b^n}{b^m} = 1 \text{ if } n = m.$$

Let's drop the positive integer requirement for exponents and look at our first exponent rule

$$b^n b^m = b^{(n+m)}$$
**Remember this one!** Eq. 1-9

and ask what we would have if $n = 0$.

The expression would be

$$b^0 b^m = b^{(0+m)} = b^m$$

Hence $b^0 = 1$.

Let's now look again at the rule

$$b^n b^m = b^{(n+m)}$$

and let $m = -n$

which leads to

$$b^n b^{-n} = b^{(n-n)}$$
$$= b^0$$
$$= 1$$

Hence

$$b^{-n} = \frac{1}{b^n}.$$

Or to write this relationship in a more general fashion,

$$\frac{b^n}{b^m} = b^{n-m}.$$

Of course, none of the denominators are zero.

We'll now delve into fractions for exponents. Once we finish with this, we can get to the topics for which you bought the book.

Let's play with the rule

$$\left(b^n\right)^m = b^{nm}$$

and replace the integer, $n$, with the fraction $\dfrac{1}{m}$, which gives us

$$\left(b^{\frac{1}{m}}\right)^m = b^{\frac{m}{m}} = b^1 = b.$$

What we have is an expression, $b^{\frac{1}{m}}$, for the $m^{th}$ root of $b$. That is, if we take $b^{\frac{1}{m}}$ to the $m^{th}$ power we get $b$. Or to write it more clearly, let $c = b^{\frac{1}{m}}$, then

$$c^m = b.$$

Hence $c$ is the $m^{th}$ root of $b$.

(If this root stuff doesn't make a lot of sense for now, don't worry. We'll dig out more roots than you want in the chapter on algebra. For now, just look at some easy examples such as $2^2 = 4$, and $2^3 = 8$, and know the 2 is the second, or square, root of 4 and the third, or cube, root of 8.)

We also use the radical sign for $b^{\frac{1}{m}}$ as

$$b^{\frac{1}{m}} = \sqrt[m]{b} .$$

Eq. 1-10

# Logarithms

Closely coupled with exponents is the mathematical operation called *logarithm*. Back in the bad old days, in the BC era[9], logarithms were useful as a computational tool[10]. Now, logarithms find limited application in some functions, in graphs where large ranges of data are plotted, and in graphs—such as the price of a company's stock— where percent changes are more important to track than actual values. "So why bother?" you may ask. Because it's there. And because there are still enough logs[11] around to build a trap for someone who doesn't understand them.

What is a log and how are logs related to exponents?

The exponent, $n$, that satisfies the equation

$$b^n = N$$

is the logarithm of $N$ to the base $b$ ($b$ is a positive number not equal to 0 or 1). And note that there is no solution for $N \leq 0$ as long as $b > 0$.

That's simple enough. Since $2^3 = 8$, then 3 is the log of 8 to the base 2.

The most common base used for logarithms is 10. It is so common that base ten logarithms are called "common logs." Another value[12] occurs in natural phenomena so often that the logs to this base are called "natural logs." We'll only use common, base 10, logs in this section, so when we write log $N = x$, the base $b = 10$ is understood.

Another example of a logarithm is

log 100 = 2, since $10^2 = 100$,

but what is log 200? That is, what exponent do we ascribe to 10 to get 200? The answer is 2.30103. How did we know that? Thankfully, someone else has worked them out and we looked it up. In the BC era we would have used published tables of logs (exciting reading, let me tell you) to find the answer. Now when you want a logarithmic value, you just go to your computer or your calculator.

---

[9] BC stands for that distant past time, "Before Computers."

[10] Some of you have no doubt heard of a slide rule. The slide rule is a mechanical logarithm calculator.

[11] Log is the abbreviation for logarithm.

[12] This value is approximately 2.71828, and is called $e$. We'll spend a lot of time with $e$ in most of the subsequent chapters.

Look at the rules for handling logarithms and see the analogy of logs to exponents. (These rules apply to any base).

Rule 1:   $\log(m \cdot n) = \log(m) + \log(n)$.                  Eq. 1-11

         Example: $\log(100 \cdot 1000) = \log(100{,}000) = 5$, and $\log(100) + \log(1000) = 2 + 3 = 5$.

Rule 2:   $\log(N^q) = q \log(N)$.

         Example: $\log(100^2) = \log(10{,}000) = 4$, and $2 \cdot \log(100) = 2 \cdot 2 = 4$.

Rule 3:   $\log\!\left(\dfrac{M}{N}\right) = \log(M) - \log(N)$.          Eq. 1-12

         Example:

$$\log\!\left(\frac{100{,}000}{100}\right) = \log(1{,}000) = 3, \text{ and}$$

$$\log\!\left(\frac{100{,}000}{100}\right) = \log(100{,}000) - \log(100) = 5 - 2 = 3$$

Rule 4:   $\log(1) = 0$.

         Combining the last two rules, we have

Rule 5:   $\log\!\left(\dfrac{1}{N}\right) = -\log(N)$;

         Example: $\log\!\left(\dfrac{1}{100}\right) = -2$, since $10^{-2} = \dfrac{1}{100}$, and $-\log(100) = -2$.

Now place yourself into the mindset of the BC folks, and you'll recognize the value of logarithms in doing messy arithmetic. Addition and subtraction are easier than multiplication and division, and multiplication is simpler than taking your number to some exponent, especially if the exponent was negative or not a whole number. If you lived in those old days and had to multiply and divide a string of numbers, you'd just look up the logarithm[13] of your numbers, do the simpler arithmetic, and then convert your final log to determine your result. Thank goodness for computers!

---

[13] The log tables only list values for numbers starting at 1 and ending <10. If you're looking for log 2, it's in the table and is equal to 0.3010. If you want log 2000, you'll just use the rule that log 2000 = log (2*1000) = log 2 + log 1000. You know that log 1000 = 3, so log 2000 = (log 2)+3 = 3.3010. Likewise, if you want log 0.002 it will be (log 2)–3 = 0.3010–3 = –2.69897. The terms in these sums have names. In the example of log 2000 = (log 2)+3, the three is called the characteristic and the log 2 (which is found in the table) is called the mantissa. You'll probably never use a log table, so this explanation is for history majors only. The log table gives the value of log 2 to only four decimal places. Your computer will do much better.

Example: If you had to work out 347*1871/21² by hand (i.e., no computer or calculator—can you imagine such misery?), you could do it. The process is not only labor intensive, but it is also fraught with the chance of careless mistakes. If you worked hard and were lucky, your calculated value is 1472.19274376417, and we bet that none of our industrious readers check our accuracy, at least without using a computer. (Yes, you're right—we used a computer.)

But we are pretending, for only this brief moment, that we have no computers. We have to go to the log tables[14] to calculate our value

$$\log\left(\frac{347 \cdot 1871}{21^2}\right) = \log(347) + \log(1871) - 2 \cdot \log(21),$$

which, after looking up the logs, becomes

$$\log\left(\frac{347 \cdot 1871}{21^2}\right) = 2.5403 + 3.2721 - 2.6444 = 3.1680.$$

The 3.1680 is the log of the number we want. We want $10^{3.1680}$ in numbers that we understand, so let's break it up and work it out.

Since $10^{3.1680} = 10^3 \cdot 10^{0.1680}$, and we know that $10^3 = 1000$, all we need is $10^{0.1680}$ for which we return to the log tables and see that (to the four places given) $10^{0.1680} = 1.472$. Hence, via logs, the solution is 1472. This errs by approximately 0.02% from the hand-cranked solution. Is that close enough? Depends on what you're doing.

## Number Bases

Most of us have ten fingers. If humans had only eight fingers we'd probably have a different numbering system. For example, computer memory chips, in their inner-most parts, can only count zero or one, so computers use base two, called binary, arithmetic.

What's all this numbering system stuff about, anyway? Let's take a look.

Recall from grade school that in our base ten system, we start to the immediate left of the decimal point and place a figure representing the number of *ones* in our total value. The second position represents the number of *tens*, the third position shows the number of *100s*, and so on. For example the number 342 tells us that we have three *hundreds*, four *tens*, and two *ones*. If we had had 342.5, we would have *five-tenths* in addition to the *hundreds*, *tens*, and *units* just listed.

To be a bit more elegant, in the base ten numbering system, each place represents ten to an integer power. We start with zero as our exponent at the immediate left of

---

[14] CRC Standard Mathematical Tables, 27th Edition, CRC Press, Boca Raton, Florida, 1981.

the decimal point. We count up as we go to the left, and we count down (negative) as we go to the right of the decimal point. Our example number, 342.5, would fill in the base ten numbering system as shown in Table 1-1.

**Table 1-1. Base ten number.**

The Value 342.5 in Base 10

| $10^3$ | $10^2$ | $10^1$ | $10^0$ | $10^{-1}$ | $10^{-2}$ |
|--------|--------|--------|--------|-----------|-----------|
| 0 | 3 | 4 | 2 | 5 | 0 |

As we mentioned, computers use base two at the CPU and RAM level[15]. Base two works just like base ten except we go in smaller steps. The first place to the left of the decimal point is still the *units* place holder. Now we can only have zero or one in that position. If we have a one in the *one's* place and need to add one more, we then have a value equal to 2 to the 1 power (that is, in base 2 when we add 1+1 we get 10). (Just as in base ten, if we had nine *ones* in that first place to the left of the decimal and added one more to it, the value in the units position would go to zero, and the position for ten to the first power would go to one.) Table 1-2 shows how we count in base two and compares the base two values to those of base ten.

**Table 1-2. The base two number system.**

BASE 2

| Exponent | 5 | 4 | 3 | 2 | 1 | 0 | −1 | −2 |
|----------|------|------|------|------|------|------|------|------|
| Base 2 Value | $2^5$ | $2^4$ | $2^3$ | $2^2$ | $2^1$ | $2^0$ | $2^{-1}$ | $2^{-2}$ |
| Base 10 Value | 32 | 16 | 8 | 4 | 2 | 1 | 0.5 | 0.25 |

If we want to express the base ten value 42.5 in base two, we must examine the number to determine how many places to the left of the decimal we need. Since 42.5 is greater than $2^5$ (32) but less than $2^6$ (64), we must place a one in the $2^5$ place. After we take care of the most significant digit, we see that we still have something left over, 42.5 − 32 = 10.5. We account for the 10.5 with a $2^3$, a $2^1$, and a $2^{-1}$ as shown in Table 1-3.

---

[15] As you probably know, computers use base 2 because it's easier to design the computer's transistors to act like switches that are either "on" or "off." These two different states naturally lead to base two. If you could figure out a way to get these same-sized transistors to work at four different states (and at the same negligible failure rate), the same-sized chip could hold four times as much information.

**Table 1-3. The base ten number 42.5 expressed in base two.**

The Base 10 Value 42.5 Expressed in Base 2

| $2^5$ | $2^4$ | $2^3$ | $2^2$ | $2^1$ | $2^0$ | $2^{-1}$ | $2^{-2}$ |
|---|---|---|---|---|---|---|---|
| 1 | 0 | 1 | 0 | 1 | 0 | 1 | 0 |

Hence, 42.5 in base ten is expressed as 101010.1 in base two.

And yes, you can do arithmetic in base two using the same rules as you used in base 10. We'll do an addition to finish this chapter.

Let's convert two base ten numbers to base two and then add them together.

The base ten numbers 42.5 and 78.75 look like 101010.10 and 1001110.11 in base 2.

Now, we'll add as we did in grade school.

    101010.10
+1001110.11
  1111001.01

Shall we check ourselves?

42.5 + 78.75 = 121.25; base 10,

and now we'll put 1111001.01 into Table 1-4, and see what it looks like.

**Table 1-4. The base ten number, 121.75, expressed in base two.**

Base 2 Table

| $2^6$ | $2^5$ | $2^4$ | $2^3$ | $2^2$ | $2^1$ | $2^0$ | $2^{-1}$ | $2^{-2}$ |
|---|---|---|---|---|---|---|---|---|
| 1 | 1 | 1 | 1 | 0 | 0 | 1 | 0 | 1 |

When we convert to base ten, we have 64 + 32 + 16 + 8 + 1 + .25, which equals 121.25.

# CHAPTER 2

# *Functions*

*With me everything turns into mathematics.* **— Descartes, 17ᵗʰ-century French philosopher**

We use the word *function* in everyday life in many different ways. Comments such as

"What is the function of that tool?"
"How do you function in that new organization?"
"The function of the armed forces is to protect the nation."

make liberal and literate use of the word function. And even from these uses of our new word-of-the-day, we get the sense that *function* denotes a state of action or of describing how something should act. Now, let's get technical.

The definition for function that is most applicable to the subject of this book is:

*A function is a correspondence, transformation, or mapping of a chosen set of variables into another set of values.*

And to continue with our fancy definitions: The first set of numbers is called the independent variable of the function and the second set is called the dependent variable[1] of the function. For any chosen value of the independent variable (any $x$ in our vernacular) there is one, and only one, corresponding value in the dependent set of values. The converse, however, is not true. Many different values of the independent variable may correspond to the same value of the dependent variable. We may refer to any related pair of values of independent and dependent variables as $(a,b)$, where $a$ is from the independent set of the function and $b$ is its corresponding value of the dependent set. In other words, when a value is independently chosen, the function prescribes a value that is dependent upon it.

Let's make a graph to demonstrate what we are talking about. However, before making the graph we need some numbers, a set of ordered pairs of numbers. As we

---

[1] The terms independent and dependent are also referred to as the domain and the range, respectively. While this terminology may have more intuitive resonance to an economic or social application, independent and dependent variables fit the engineering use of functions. The range could be thought of as the scope or range of possible solutions that can be mapped using the domain. And as long as we're on terminology, the independent variable, when written within the function, is often called the **argument** of the function.

go along we'll let mathematical formulas help generate those pairs, but for now we'll somewhat arbitrarily generate a set to graph. We show our pairs in Table 2-1 and graph them in Figure 2-1. This not an exciting plot, and it has no physical significance. It does, however, demonstrate some of the salient features of functions such as: 1) There is only one value of the dependent variable for each value of the independent variable. 2) A given value of the dependent variable may correspond to more than one value of the independent variable, such as the sets (–2,2) and (0,2). 3) Although we normally plot the independent variable as it monotonically increases, the dependent variable goes up and down in value.

**Table 2-1. Eleven ordered sets of values.**

| Independent Variable | Dependent Variable |
|---|---|
| –5 | –3 |
| –4 | –4 |
| –3 | 0 |
| –2 | 2 |
| –1 | 0 |
| 0 | 2 |
| 1 | 1 |
| 2 | 2 |
| 3 | 3 |
| 4 | 4 |
| 5 | 3 |

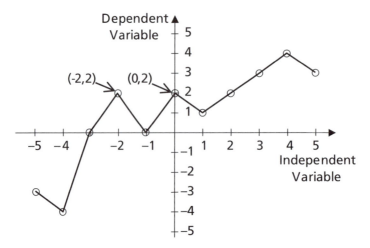

**Figure 2-1. A plot of the ordered sets of values from Table 2-1.**

How are we sure that the dependent variables are truly dependent upon the independent values, as prescribed by the function? Correlation does not necessarily mean causation. In fact, there may be more than one independent variable affecting the dependent variable. For example, we could develop a function for the volume of concrete needed to build a bridge over a river. While the distance the bridge must span is certainly an important variable, other independent variables such as service load, channel depth, wind load, and type of bridge certainly play a significant factor. The function may have more than one independent variable. More on this topic later.

Nothing in this definition of a function demands a mathematical relationship between the independent and the dependent variables. If we list the gross domestic product (GDP) of the nations of the world, we'd have a function where the nations are the independent variable and their wealth is the dependent variable. At any given time, each nation (from the independent variable set) has only one number for its GDP (the value of the dependent variable) although more than one nation may have the same GDP. OK, enough economics, let's get back to math and functions where the dependent variable is related to the independent by a mathematical relationship.

From here on when we say "function" we'll mean mathematical function, where the dependent variable is related to the independent variable by a mathematical operation. We'll normally name the function $f$ where $f$ takes a value from the independent set of variables and transforms it into a value in the dependent set. We'll indicate this by

$f(x)$ = whatever operations we want, where $x$ is the only variable.

If there is more than one variable[2] that can affect the value of $f$, we'll write it as $f(x,y,z,...)$.

As we talk about different functions we'll call most of them $f(x)$ as long as they are separate and independent of each other. As we put some physics to these functions, $x$ might be position, time, angle, or mass, and $f(x)$ might be velocity, force, or weight. Sometimes we'll change the $x$ and $f(x)$ designation to something that helps as a mnemonic, such as using $t$ for time or $v(t)$ if we're talking about velocity as a function of time. Also, if we have two functions that are related, such as the weight of a ship and the volume of water it displaces when we're writing these two functions as dependent on the number of cargo boxes brought aboard, then we'd certainly use two different designations, such as $f(x)$ and $g(x)$ or $w(x)$ and $v(x)$ (or we could use $n$ instead of $x$ for the number of boxes). It doesn't matter what we call them as long as we are clear and consistent.

Let's look at some functions.

---

[2]  Even nonmathematical functions can have more than one variable. For example, we can look at the nation's GDPs as a function of both nation and time.

$f(x) = x$

is as simple as functions get. For every value of $x$ there is a corresponding value $f(x)$ that is equal to that chosen value of $x$. We can make a table of these values. We can plot values from that table in a graph where we put values of $x$ on the horizontal axis and the corresponding values of $f(x)$ on the perpendicular axis. Where the two intersect place a "data point" designating the pair of numbers. Or we can leave the function in its mathematical form and do all sorts of neat things with it as you'll see in subsequent chapters. For now, we'll just leave that function alone and look at some other examples of functions.

Another function is

$f(x) = 2$.

"Wait a minute," you might say. "There isn't even an '$x$' in this thing." And you're right. But it is still a function. For every value of $x$ there is a corresponding value of $f(x)$ and that value is 2, regardless of what $x$ may be.

Let's continue with

$$f(x) = -2 + 3x^2$$

Here we not only have a coefficient in front of $x$ and an additive constant, but we're also raising $x$ to a power. Let's make a table and graph this relationship to see how $x$ is related to $f(x)$.

We may also write these values as $(x, f(x))$. For example $(-3, 25)$ refers to the first row in Table 2-2.

**Table 2-2. *x* is the independent variable and *f(x)* is the dependent variable.**

| $x$ | $f(x)$ |
|-----|--------|
| −3  | 25     |
| −2  | 10     |
| −1  | 1      |
| 0   | −2     |
| 1   | 1      |
| 2   | 10     |
| 3   | 25     |

Of course, we could have chosen any values of $x$ we wanted and then calculated the corresponding value of $f(x)$, but we're keeping the example close to home where the interesting stuff is happening. The most important thing to note is that both one and minus one produce the same value of $f(x)$. You can have only one value of $f(x)$ for each value of $x$, but there are no fundamental restrictions on how many $x$'s can produce the same value of $f(x)$. It all depends on the function. Next, and we hope that this is obvious, $x$ as well as $f(x)$ can have both positive and negative values. And finally, for this function, there are no restrictions on $x$; it can be any finite value.

Some functions have restrictions on the permissible values of $x$. The function

$f(x) = 3/x$

is a valid function, but it cannot take on all values. It gets very messy when $x$ approaches zero. So, for this deceptive function, the independent variables are all numbers except zero. And as we get into functions that describe physical phenomena, we must be reasonable with our $x$ values. If we have a function for the velocity, as a function of time, of a falling cannonball, we can only calculate that velocity from the time it starts to fall until it hits the ground.

## Plotting Functions

We've already graphed a function, and we guess that graphing is old-hat to most of you, but let's get formal on graphs for a few paragraphs. First, the formal definition of a graph:

> A graph of a function is the visual representation of a function, and it consists of all points, $x$ and $f(x)$, where $x$ is in the independent variable set of the function and $f(x)$ is the corresponding value in the dependent variable set.

While we're on this tack, let's define a few more things. In this chapter we'll stay with the **Cartesian**[3] **coordinate system**[4]. This is the basic system of two perpendicular, intersecting[5] lines that usually intersect at the origin (the point (0,0)) of their respective **axes**. We normally draw these axes as a horizontal line and a vertical line

---

[3] The Cartesian system honors Rene Descartes, the seventeenth-century philosopher, scientist, and mathematician. He is most widely known for his maxim, "I think, therefore I am," which we don't understand. We know many folks who have no thoughts whatsoever, but who quite vividly exist. But we think that Rene's coordinates are good stuff.

[4] We'll get to other coordinate systems in later chapters that are designed for things that look like cylinders and spheres.

[5] To say that two lines are perpendicular and that they intersect is not redundant. Two nonparallel lines will intersect only if they lie in the same plane. Conversely, two intersecting lines define a plane.

(though any starting angle will work as long as the other axis is perpendicular[6] to it). The horizontal line usually (and always in this book) represents the values of the independent variables, or $x$ values, and it is called the **abscissa** or **x-axis.** The perpendicular axis is called the **ordinate** or **y-axis**. Since $f(x)$ gives us a value on the $y$-axis, $f(x)$ is often referred to as[7] $y$.

Since these crossing lines form four sections, we call each section a quadrant and we label the quadrants 1 through 4 as shown in Figure 2-2. From this definition we know the names of the proper quadrants for points such as (2,2), (–2,2), (–2,–2), and (2,–2). Pairs of numbers such as these, which give the location of a point on a graph, are called **coordinates** of the point.

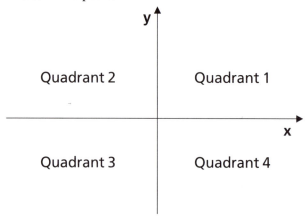

**Figure 2-2. Quadrants of a graph.**

Now that we have four coordinate points on a graph, what can we do with them? We can determine the distance between any two points, the slope (which we'll soon define) between them, and what sort of function, if any, may have generated these points.

## Distance

The distance between any two points is simply how far you'd have to go if you went in a straight line from one point to the other one, if both axes have the same units. If the $x$-axis is in meters and the $y$-axis is in kilometers, some conversion would be in order. If the $x$-axis is in time and the $y$-axis is in velocity, then the concept of distance would have little meaning without a lot of conversion.

---

[6]  The two axes do not have to be perpendicular. To draw them otherwise is not productive, and we know of no reason to even consider nonperpendicular axes.

[7]  Always look before you leap, especially when leaping into mathematical symbology. Other functions may be dependent upon two (or more) variables, and the writer may choose $y$ as the second variable. If the author does not adequately define the terminology and symbols used in the work, get another book.

The distance between the coordinates (–2,–1) and (2,–1) is easy to determine. You start at (–2,–1), walk to the right for two units to (0,–1), and then walk two more to (2,–1). You traveled four units, hence the distance is four. Now, we bet that you can figure out how far it is when you go from (2,–1) to (2,2).

Now you are sitting at (2,2). You have walked a total of seven units to get there. If you want to go directly back to (–2,–1), can you take a short cut? Of course you can. You know that the shortest distance between any two points is a straight line, and thanks to Mr. Pythagoras[8] you can calculate this distance. The *Pythagorean theorem* states that the distance between the two extreme points on a right triangle is the square root of the sum of the shorter two sides squared. That sounds like a mouthful, but it isn't. The three coordinates, (–2–1), (2,–1), and (2,2) form a right triangle. Since you've walked the two legs, you know those distances, so the straight line distance from (2,2) to (–2,–1) is

$$D = \sqrt{4^2 + 3^2}$$
$$= 5 .$$

We plot your trip in Figure 2-3.

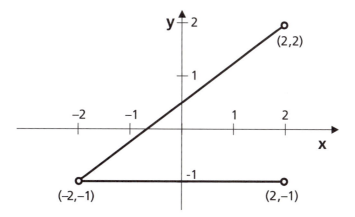

**Figure 2-3. Pythagorean theorem in action.**

Now, does this give you a hint for finding the distance between any two points on a graph? We hope so. If you have any two coordinate points on a graph, say (–1,1) and (1,7) and want the distance between them, just imagine a third point at (1,1) and you have a right triangle. The horizontal leg is (1–(–1)) units long and the vertical leg is (7–1) units long. The distance between your two points is

---

[8] The Greek mathematician and philosopher, Pythagoras, lived some 2500 years ago. He is often considered the first mathematician.

$$D = \sqrt{\left(1-(-1)\right)^2 + \left(7-1\right)^2}$$
$$= 2\sqrt{10}$$

Does this point to a general distance equation? It's

$$D = \sqrt{\left(x_2 - x_1\right)^2 + \left(y_2 - y_1\right)^2}$$

where $(x_1,y_1)$ and $(x_2,y_2)$ are any two points on the graph. This equation even works for such no-brainers as the distance between (2,2) and (–2,2). Try it.

## Slope

It's time for a break. When we hear the word slope we immediately think of ski slopes, which is a good example of a slope (the best, we think). A slope is the change in the height of something as you change position. A slope in this math book need not involve mountains but might refer to the change in your car's velocity as you increase the engine size. It's the change in the dependent variable as you change the independent variable. We will use $m$ to designate the slope. We have no idea why $m$ is given this honor, but that's the language, so we use it.

In a mathematical definition the slope between two points is

$$m = \frac{y_2 - y_1}{x_2 - x_1}$$

which fits the common conception that the slope is the *rise*, $y_2 - y_1$, divided by the *run*, $x_2 - x_1$.

An example is the slope between (–2,–1) and (2,2). The difference in $y$ is 3, and the difference in $x$ is 4. Thus, the slope is 0.75. And yes, it's +0.75, not –0.75. We know that if it's uphill from your house to school then it's downhill[9] from school to home, but we'll always determine the sign of the slope starting from the lower $x$ value. If you, in conversation, wish to state it otherwise, do it. Just be careful of your terminology.

Please note that $m$ is the slope between two points. The slope equation says nothing about what kind of function might connect those points. If, in the real world, we wanted to know the slope between Denver and Salt Lake City, this equation would ignore the intervening Rocky Mountains. We'll look at points, slopes, and functions in the next section and put some of this stuff together.

## Functions and Coordinate Points

Let's take the four ordered sets of values (–2,–2), (–2,2), (2,–2), and (2,2), and plot them as points on a graph. Now, ask yourself what functions could generate these points? If we take all four sets of points, any three sets of them, the set (2,2) and (2,–2),

---

[9] Yes, we know that your grandparents said it was uphill both ways.

or the set (–2,–2) and (–2,2), the answer is easy: there are no functions that can pass through those selective sets of points. Why? Because each of those sets of points contain at least one value of *x* that corresponds to two values of y, and that's been a "no-no" since the beginning of the chapter.

Of the pairs of points, only [(–2,2) and (2,2)], [(–2,–2) and (2,–2)], [(–2,–2) and (2,2)], and [(–2,2) and (2,–2)] may represent functions (see Figure 2-4). "What functions?" you ask. Many functions, but we'll use Occam's razor[10] to shave away the excess and just take the simplest functions that incorporate these sets of points.

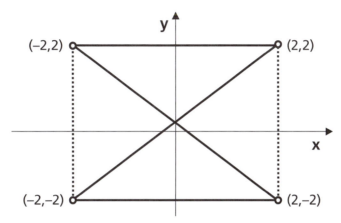

**Figure 2-4. Allowed sets of points that may be represented by a function.**

Each of the two sets [(–2,2) and (2,2)] and [(–2,–2) and (2,–2)] has the same value for y, hence the slope between these points is zero. These two sets of points are satisfied by

$$y = 2$$

and

$$y = -2$$

respectively.

---

[10] Occam's razor—from William of Occam, 14th-century English philosopher—is a rule in science and philosophy interpreted to mean that the simplest of two or more competing theories is preferable and that an explanation for unknown phenomena should first be attempted in terms of what is already known. In other words *keep it simple*. Occam's razor is worth remembering and applying.

The other two sets of points, [(–2,–2) and (2,2)] and [(–2,2) and (2,–2)], have nonzero slopes between them. So we'll use that razor to take the next most interesting function: a straight line with a nonzero slope. We've already calculated the slope between (–2,–1) and (2,2). Run the numbers through the slope equation for these two sets of points and see that the slopes are +1 and –1.

So, functions that satisfy these sets of points are

$$f(x) = x$$

and

$$f(x) = -x$$

respectively (see Figure 2-4).

These simple functions where $f(x) = x$ or $f(x) = -x$ are the simplest examples of **linear equations**.

By linear, we mean that some given change in the independent variable will produce the same change in the dependent value regardless of where this change takes place. For example, the function, $f(x) = x$, increases by one unit when $x$ increases from 0 to 1. It also increases by one unit when $x$ increases from 100 to 101. Try this with $f(x) = x^2$ and see the difference. If this discussion seems a bit trite to you, just hang on. You'll see a lot more of linear equations as we go on.

Let's go one step further. Let's move the set of points, [(–2,–2) and (2,2)], vertically by three units in the $y$ direction. They become the points (–2,1) and (2,5). Do the slope calculation, and you'll see that it's still 1. You still have a linear equation (two nonvertical points can always be described by a linear equation), but you'll also see that $f(x) = x$ is no longer the describing function. You are missing the $y$-axis intercept, the value of your function when $x = 0$. When your line goes through the coordinate (0,0) you can write your linear equation as

$$f(x) = mx$$

where $m$ is the slope of the line ($m = 1$ in the previous examples).

Now you need the full linear equation

$$f(x) = b + mx,$$

where $b$ is the value of $f(0)$. That is, $b$ is the value of the function when it intercepts the $y$-axis.

How can we determine $b$? Well, you can plot the two points on a graph and see where the function crosses the $y$-axis. That's not very accurate, but it will give a close estimate. In this example, your first point was at (–2,1). If you proceed to the right (the positive $x$ direction) two units, your $x$-coordinate becomes zero. Since your slope is one, the *y*-coordinate also increases by two units to become $1 + 2 = 3$. Hence, your **y-intercept** is 3. So *b* is equal to 3. We already know that *m* is 1, so your equation is

$$f(x) = 3 + x.$$

Is there a better way to calculate *b*?

Sure there is. Two points are all you need to define a linear equation, so consider the two points $(x_1, y_1)$ and $(x_2, y_2)$. You already know how to determine the slope, *m*, from the equation

$$m = \frac{y_2 - y_1}{x_2 - x_1}.$$

Since the slope equation works for any two points on the linear curve, it works between $(x_1, y_1)$ and $(0, b)$ where we have taken the second point as $x = 0$ and $y = b$. We now write the slope equation as

$$m = \frac{y_2 - b}{x_2 - 0}$$
$$= \frac{y_2 - b}{x_2}.$$

We now have one equation and one unknown[11], so we solve for *b* as

$$b = y_2 - mx_2.$$

We have an assignment for you. On a sheet of graph paper, make two marks at any two nonvertical positions. Determine the $(x_1, y_1)$ and $(x_2, y_2)$ positions of those points and calculate *m* and *b* from the equations that we've just covered. Now, use a straightedge to connect your two points with the *y*-axis. Hopefully, your straightedge will cross the *y*-axis at your calculated value of *b*.

Going to the other extreme, we can fit a multitude of functions to any single point on the graph. For example, any linear function, $f(x) = b + mx$, will go through the point (2,2) just as long as $b = 2 - 2m$. You just choose any finite value for *m* and the concomitant value of *b* is readily calculated. But don't do it. Mr. Occam would advise extreme caution in fitting any function to a single point.

> Note that $f(x) = b + mx$ can also be written as $y - mx - b = 0$, where we use *y* for the value of the dependent variable, *f(x)*. Since you can multiply both sides of the equation by any nonzero term, we can write the equation of the line as $Ay + Bx + C = 0$. The simplest values for *A*, *B*, and *C* are 1, −*m*, and −*b*, respectively. We'll do more with this form in *Matrices*, Chapter 3.

---

[11] If this is a new concept to you, we beg your indulgence for its early introduction. We'll go into a lot more detail in the chapters on algebra and matrices.

**Examples:**

(1) You have two equations, $f(x) = 5 + 3x$ and $-10y + 30x + 50 = 0$. Do these equations represent the same line?

You can get the answer in different ways.

(a) Plot each and see if they fall on top of each other.

(b) Make a table and calculate *f(x)* and *y* for two different values of *x*. If *f(x)* and *y* have the same values for the same *x*, then they are the same equation. Be warned: this only works for linear equations and be sure to use more than one data point.

(c) Take the second equation, $-10y + 30x + 50 = 0$, put the *x* term and the constant on the righthand side, and reduce the *y* coefficient to unity. Do the equations look the same?

And, yes, they are the same function.

(2) Do the lines represented by the equations $f(x) = 5 + 3x$ and $g(x) = 3 + 2x$ ever cross? If they do, where? Can you change *b* of either equation to keep them from crossing? Can you change a value of *m* to keep them from crossing? If so, how?

Let's take these suggestions one at a time. Figure 2-5 will help clarify the answers.

(a) Do the lines represented by the equations $f(x) = 5 + 3x$ and $g(x) = 3 + 2x$ ever cross?

Yes, they cross. Both are straight lines with different slopes. They must cross somewhere.

(b) If they do, where?

All you need to do is to ask if there is a value of *x*, which we'll call $x'$, where $f(x') = g(x')$ or where $5 + 3x' = 3 + 2x'$. This equation is satisfied when $x' = -2$, and at $x' = -2$, $f(-2) = g(-2) = -1$. Hence, the two lines cross at $(-2, -1)$.

(c) Can you change "*b*" of either equation to keep them from crossing?

No. If you change *b* in either (or both) equations, the lines will cross at a different point. But they will cross.

(d) Can you change a value of *m* to keep them from crossing? If so, how?

If two lines have the same slope, they are either *the same line*, or they are parallel lines. The first answer is too metaphysical for us engineering types, so we'll concentrate on the second. Even though railroad tracks seem to converge together in the distance, parallel lines do not ever cross.

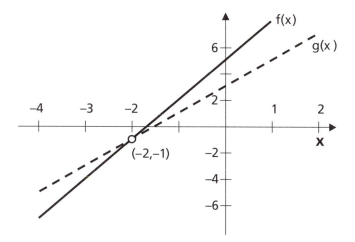

**Figure 2-5. Examples of lines generated by linear equations.**

## Translation of Coordinates

Before going on to fancier functions, let's look at something that you may say is obvious. There will be a time when you want a fresh start and you may want to plot your data on new axes. In a prior example we had coordinate points $(-2,-1)$ and $(2,2)$. If now we decide to make the first point the origin, we can do it. You are only shifting (or translating) the coordinates by $-2$ units in $x$ and 1 unit in $y$. And your new coordinate system $(X,Y)$ is related to the old $(x,y)$ by

$$Y = y + 1$$

and

$$X = x + 2.$$

Solve these for $y$ and $x$ sustitute, and our linear equation

$$y = 0.5 + 0.75x$$

becomes

$$Y - 1 = 0.5 + 0.75(X - 2)$$
$$Y = 0.75X$$

These two renditions of the same function on two different coordinate systems are shown in Figure 2-6.

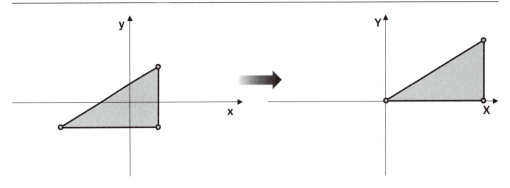

**Figure 2-6. Plot of the same function on two different coordinate systems.**

## Other Functions

The title of this section is "other" functions, not "all other" functions. There is always another function lurking out there somewhere. We'll only look at a couple of other functions now and get to many others in subsequent chapters.

For starters, let's look at

$$f(x) = b + ax^2 .$$

We've already plotted an example of this form,

$$f(x) = -2 + 3x^2 ,$$

so there's not much new to say except:

1) This is not a linear function. It has a square term in it, so the increase in *f(x)*, for a given increase in *x*, is dependent on where that *x* happens to be.

2) If you look at the plotted curve[12] of the equation you readily see that the tangential[13] slope is continuously changing.

   and

---

[12] We tend to call all plotted equations curves even if they are straight lines. This curve, of course, is not straight.

[13] A curve drawn from a linear equation has an unchanging slope equal to *m* at all points. A curve drawn from an equation such as $f(x) = b + ax^2$ has a different slope at every value of the independent variable. We can estimate the slope of this curve at any given point by drawing a straight line (called a tangential line) that touches but does not cross our curve. The slope of the straight line is the slope of the curve at that point. This idea is simplified when we get to calculus.

3) Though you can still calculate the straight-line distance between any two points on this curve, you cannot use that formula to determine the length along the curve.

These three characteristics are shown in Figure 2-7.

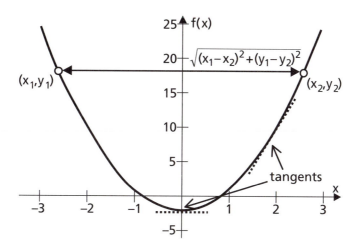

**Figure 2-7. Characteristics of a nonlinear curve.**

# Log Functions

Let's look at the value[14] of a share of Microsoft stock over a seven-year period. The value is listed in six-month steps in Table 2-3.

**Table 2-3. Value of a share of Microsoft stock over a seven-year period.**

| Period # | Value |
|:--------:|:-----:|
| 1 | $5 |
| 2 | $15 |
| 3 | $15 |
| 4 | $20 |
| 5 | $30 |
| 6 | $35 |
| 7 | $35 |

---

[14] Adjusted for stock splits. (If you don't know what that means, just wait until you're a math expert and start investing your new wealth—you'll quickly learn.)

*Table 2-3 (continued)*

| | |
|---|---|
| 8 | $60 |
| 9 | $80 |
| 10 | $95 |
| 11 | $125 |
| 12 | $85 |
| 13 | $40 |
| 14 | $70 |

As you see, the price has varied greatly, and if we plotted dollars vs. time period on a linear graph (Figure 2-8), the price changes in its early days are just a wiggle, but the percentage change (which is the important parameter to an investor) might be large.

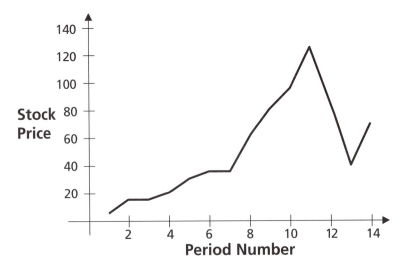

Figure 2-8. Stock price vs. time period using linear axes.

Let's take the logarithm of the share price and plot that against time (Figure 2-9). Since logs work with exponents, we see that a factor of two change in the early, low-cost days shows up just as visually as the same percentage change at a later time.

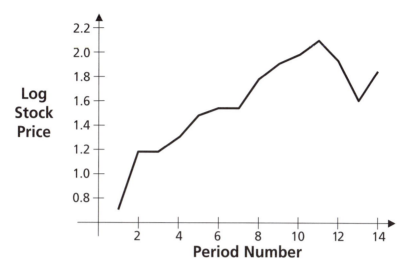

**Figure 2-9. Log of stock price vs. time period.**

Looking at log values, however, is not very elucidating, so what to do? We need a *y*-axis that is logarithmic rather than linear. The data is plotted using a logarithmic vertical axis in Figure 2-10, and we readily see that the largest relative gains occurred in the earlier time periods.

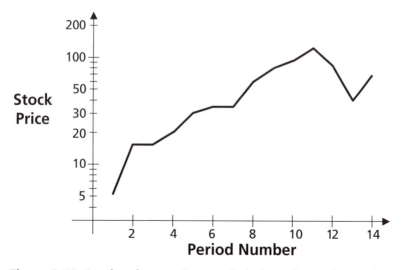

**Figure 2-10. Stock value vs. time period plotted on a log scale.**

# *Algebra*

*Algebra is the intellectual instrument which has been created for rendering clear the quantitative aspects of the world.* — **Alfred North Whitehead, British mathematician**

Now we delve into the unknown. We seek, we find, and we do not yield until we fulfill our quest to lay bare a numeric value for the baffling $x$. It's a mystery why $x$ was chosen as the great unknown. It is part of our language, however, as the identifier of mystery and intrigue. Look at popular culture and all that is called $x$. X-rays, well known today, gained their name when Roentgen[1] was working with electron beams late in the nineteenth century. He knew that some sort of ray was sneaking around and fogging his film, but he had no idea what it was, so he called it "x-ray."

Algebra is so fundamental to engineering, it's almost like breathing. So let's get on with it, beginning very simply.

Let's say that William and his colleague Jacob have six computers. If they share equally (and that's a big if), how many computers will each have?

Let $x$ represent the number of computers each engineer has. We have two engineers and six computers, so

$$2x = 6.$$

As we saw in Chapter 1, we can divide each side by 2, so,

$$x = 3.$$

That is, each engineer has three computers. (Although we doubt it—with the William and Jacob we know, William would have four computers and a lawsuit, and Jacob would have only one computer after losing the other one throwing it at William. But this is math, not sociology, so we can assume a perfect world, designed to our liking.)

---

[1]  Wilhelm Conrad Roentgen, German physicist, discovered x-rays by accident in 1895.

The symbol $x$ is going to represent all sorts of things. Later, we'll couple equations with more than one unknown. In those examples $x$ in one equation will have the same value (or set of values) in each of the coupled equations. You'll see more of this in Matrices, Chapter 4.

Let's write the last example in a slightly different way[2]. Let's subtract 6 from each side and get

$$(2x - 6) = 0.$$

If the righthand side of the equation is zero, then the lefthand side must be zero as well. The value of $x$ needed to make the lefthand side equal to zero is $x = 3$, since 2 * 3 is 6, and 6 minus 6 is zero, and all is well. (Conversely, divide both sides by 2. That gives you $(x - 3) = 0$. By casual inspection you see that letting $x = 3$ gives the desired result.) We call the value of $x$ that makes the equation equal to zero the ***root***[3] of the equation.

Now, let's take the next step and put some more brackets (with numbers and variables) into the equations. Let[4]

$$(2x - 6)(3x + 6) = 0.$$

Let's think now. If the stuff in either set of parentheses equals zero then we have at least partially solved the problem. Regardless what the other parentheses equals—ignoring infinity[5] for the time being—zero times any number still equals zero. (We say "partially solved" for we want to find all values of $x$ that satisfy the criteria that the product of the values in parentheses equals zero. We want to find all of the roots.)

Just take the sets of stuff in each of the parentheses one at a time, and you easily solve for two different values of $x$.

$x = 3$ for the first parentheses, and

$x = -2$ for the second.

---

[2]  The parentheses are not necessary in this equation. We like to use them for clarity.
[3]  Don't confuse *root* of an equation with the *square root* of a number.
[4]  Here, of course, the parentheses are necessary.
[5]  Essentially, we'll ignore infinity throughout the book.

To make life a bit more interesting, multiply the terms in the two sets of parentheses above. We think you know how to do this, but let's walk through it. Starting with

$$(2x-6)(3x+6)=0$$

we multiply each of the terms in the second parenthesis by the terms in the first parenthesis one at a time as

$$2x(3x+6)-6(3x+6)=0$$

to get

$$6x^2+12x-18x-36=0.$$

We now collect terms that have a common power of $x$,

$$6x^2-6x-36=0.$$

(Of course $6x^2-6x-36=0$ can be reduced to $x^2-x-6=0$, but allow us to keep the nonreduced form in these examples.)

Although this came from the very same set of $x$'s and coefficients in our double parenthesis, most of us are not smart enough to look at this form and see that $x$ equaling three and minus two will produce the desired result of bringing the equation to zero. Of course, you can substitute three and then minus two for $x$ and demonstrate that the equation goes to zero in each case. Also, try as you might, you will find no other value for $x$ that will solve the equation.

So, looking at this rendition of the equation, what do we do? How do we determine the values of $x$ that make the equation equal zero?

There are ways to solve for the values of $x$, but now is a good time to pause and talk about where we're going and what algebra is all about.

Algebra is the study of mathematics in which the operations and procedures of addition and multiplication are applied to variables rather than just to specific numbers. (And yes, we can divide also, but division can be messy. We'll get to division later on.) In this section we'll study the algebra of polynomials of one variable.

To start, let's settle on a definition for polynomial:

$$2x-6$$
$$6x^2-6x-36,$$

and

$$x^4+3x^2-1$$

are polynomials.

Even the number 2, all by itself, is a polynomial even though it's a single number.

Examples of what isn't a polynomial follow:

$\sqrt{x} + 4$,

$x^{-3} + 2$,

and

$2^x - 5$.

So let's clarify. A polynomial is a number described as

$$ax^n + bx^{n-1} + cx^{n-2} + \cdots + gx + h$$

where $n$ and all other powers of $x$ are positive integers or zero, and

$a, b, ..., h$ are numbers, real or imaginary.

If $a$ in the above example is not zero, then the polynomial is of degree[6] $n$. This holds even if all the other coefficients are zero.

So

$$6x^2 - 6x - 36$$

is a second-degree polynomial and

$$2x = 6$$

is a first-degree polynomial.

What we were doing at the beginning of the chapter with

$$2x = 6$$

and

$$6x^2 - 6x - 36 = 0$$

was looking for the zeros—the roots—of those polynomials. And when we looked at that second equation as

$$6x^2 - 6x - 36 = (2x - 6)(3x + 6),$$

we were factoring the equation. And after factoring, we can readily identify the roots of the polynomial.

We'll do a lot more searching for these zeros and factoring later, but first let's look at the arithmetic of polynomials.

---

[6] Some texts use the terms *degree* and *order* interchangeably. We are saving *order* for its special use in differential equations.

Once you prescribe a value for $x$, the polynomial is just a number, so it seems obvious that polynomials follow the rules of arithmetic.

Want to add them? Sure you can. Just don't mix the powers of $x$. For example, if you add

$$3x^3 + 5x - 4$$

to

$$2x^4 + 7x^2 + 9$$

you get

$$5x^3 + 7x^2 + 5x + 5.$$

You had a $x^3$ term in each polynomial, only one nonzero $x^2$ term, one $x$ term, and two constants. The powers of $x$ were kept together when we combined terms. Note that with addition you may lose terms of a given power of $x$. You may lower the degree of the polynomial if the highest degree terms cancel each other, but you'll never gain a power you didn't start with. You also will never raise the degree of the polynomial doing addition.

Pick any value of $x$ and plug it into the equations. Add the first two equations and see that the sum equals the third. Just be sure to use the same value of $x$ in all three equations.

Since substitution is an obvious extension of addition, we won't insult you by providing an example.

We have already shown a simple multiplication in the introductory portion of this chapter. To multiply two polynomials, just multiply all terms in the second polynomial by each term in the first and then collect terms of like powers of $x$. Note that the degree of the product is the sum of the degrees of the two multipliers. You may lose some lower degrees of $x$ as you combine terms, but the degree of your new polynomial will always increase unless you have only zero degree terms in one or both of the multipliers.

The division of polynomials is messy. For starters, you can't divide a polynomial by a higher degree polynomial. Sure, they are both numbers, but look at the simple example of

$$\frac{x^4}{x^6}.$$

You know how to handle powers and the quotient (as discussed in Chapter 1) is

$$x^{-2}.$$

It's a number, but it's not a polynomial by our narrow-minded definition of polynomials.

Let's do a simple example and then move on to more interesting ones. We'll start with one where you already know the answer. Remember the polynomial

$$6x^2 - 6x - 36 = (2x - 6)(3x + 6) ?$$

We'll divide $6x^2 - 6x - 36$ by $(2x - 6)$ and see if it equals $(3x + 6)$. We proceed just as we would with simple arithmetic where

$$
\require{enclose}
\begin{array}{r}
3x \phantom{xxxxxxxxx} \\
(2x-6)\enclose{longdiv}{6x^2 - 6x - 36} \\
\underline{6x^2 - 18x \phantom{xxxx}} \\
-12x - 36
\end{array}
$$

is the first step. The second step, just as in simple arithmetic, is to see if you can divide again as

$$
\begin{array}{r}
3x + 6 \phantom{xxxxx} \\
(2x-6)\enclose{longdiv}{6x^2 - 6x - 36} \\
\underline{6x^2 - 18x \phantom{xxxx}} \\
12x - 36 \\
\underline{12x - 36} \\
0 + 0
\end{array}
$$

Since $6x^2 - 6x - 36$ can be factored as $(2x - 6)(3x + 6)$, you could have performed the division as

$$\frac{6x^2 - 6x - 36}{(2x - 6)} = \frac{(2x - 6)(3x + 6)}{(2x - 6)} = (3x + 6)$$

since the $(2x - 6)$ terms cancel each other.

We'll do one more division, without factoring, before we move along. Let's divide $2x^4 - x^3 + 7x^2 - 2x - 2$ by $2x^2 - x + 1$, which is

$$
\begin{array}{r}
x^2 + 3 \phantom{xxxxxxxx} \\
(2x^2 - x + 1)\enclose{longdiv}{2x^4 - x^3 + 7x^2 - 2x - 2} \\
\underline{2x^4 - x^3 + x^2 \phantom{xxxxxxx}} \\
6x^2 - 2x - 2 \;. \\
\underline{6x^2 - 3x + 3} \\
x - 5
\end{array}
$$

The answer is $x^2 + 3$ with remainder $R = x - 5$.

As you might have expected, you lose degrees in your answer with division unless the divisor is of zero degree. That is, it's just a constant.

Note: Let's back up and think for a moment about some of the basics of doing arithmetic with polynomials and compare it to doing arithmetic with numbers expressed in powers of ten.

You wouldn't try to add (or subtract)

$2.01 \times 10^2$ and $3.479 \times 10^3$

without shifting, at least in your head, the decimal place with the concomitant change in the exponent. You would convert

$2.01 \times 10^2 + 3.479 \times 10^3$

to

$2.01 \times 10^2 + 34.79 \times 10^2$

which, of course, equals

$36.8 \times 10^2$ or $3.68 \times 10^3$.

The same procedure holds with polynomials except that you can't readily shift the decimal places and powers when it's $x$ instead of the number ten that is raised to different powers. You can only combine like powers.

Multiplication and division of polynomials are analogous to operations with powers of ten. They're so similar that we'll let you sort them out for yourself.

## First-degree Polynomials

Remember the simple computation that we used at the beginning of the chapter,

$2x - 6 = 0$ ?

Instead of looking for the root of this first-degree polynomial—that is, only determining when it equals zero—let's let $x$ vary. Doing so, we have a first-degree function[7]. Let's designate these different values of the polynomial as

$y(x) = 2x - 6$

where *y(x)* is that value determined by multiplying the value of *x* by 2 and then subtracting 6 from that product, such as shown in Table 3-1.

---

[7] You may want to review Chapter 2 at this time.

**Table 3-1. Values of a linear function.**

| x | y(x) |
|---|---|
| −4 | −14 |
| −3 | −12 |
| −2 | −10 |
| −1 | −8 |
| 0 | −6 |
| 1 | −4 |
| 2 | −2 |
| 3 | 0 |
| 4 | 2 |

Of course, there is no requirement to restrict the values of $x$ to integers.

Here we have created an **ordered set** of **pairs of numbers**. To each number in the first column we have a determined value for the second column. And for whatever value we independently choose for the first column, the corresponding value in the second column is quite dependently set at twice the value of the first minus six. Of course, we can choose other polynomials or other functions for the dependent variable, and needless to say, we'd get different values.

This example is a simple linear equation, or linear function, like those we plotted in the chapter on functions. Remember the term *linear*. Linear means that for every unit increase of the independent value, $x$, we get the same increase in the value of the dependent variable, $y(x)$, regardless of where we are on the domain of the independent variable. Or to say all that in simple English: any single step you take in $x$ gives you the same increase in $y(x)$ regardless where you are on the $x$ axis.

We can write a general expression for a linear equation as

$$y(x) = mx + b$$    **Remember this one!**  Eq. 3-1

which describes a graphed (**Figure 3-1**) function where

$m$ is the amount that $y(x)$ increases for each unit change of $x$

and

$b$ is the value of $y$ when $x = 0$. That is, $y(0) = b$.

Just for the fun of it, note that the zero value of the function, that is $y = 0$, occurs at

$$x = {-b}/{m}.$$    Eq. 3-2

To summarize: the graphed linear function crosses the $y$ axis at

$x = 0$ (of course)

and

$$y(0) = b\,,$$

and the line crosses the x axis at

$$x = {}^{-b}\!/_{m}$$

and

$$y({}^{-b}\!/_{m}) = 0 \text{ (of course)}.$$

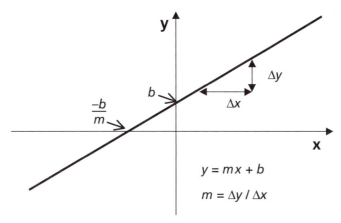

**Figure 3-1. Graph of a linear function showing axis crossings.**

There's one more manipulation to show. Take the general form of the linear function

$$y(x) = mx + b\,.$$

Put all terms on the lefthand side as

$$y(x) - mx - b = 0\,.$$

Multiply through by any number you choose and write *y(x)* as *y*, and you have

$$ax + cy + d = 0$$

where

$$m = {}^{-a}\!/_{c}$$

and

$$b = {}^{-d}\!/_{c}\,.$$

Remember that the *x* and *y* are ordered pairs described by the linear polynomial and that they fall on the graph of the function.

Note that we placed no restrictions on *a, c,* or *d* (except for the tacit assumption[8] that *c* does not equal zero). What we have said is that any three numbers, as long as $c \neq 0$, written as

$$ax + cy + d = 0$$

describe a linear function. We'll make more use of this form in Chapter 4, Matrices, when we put equations together to solve multivariable[9] problems.

## Second-degree Polynomials[10]

If we add another linear polynomial to the one that we were using before, we still have a linear function. Now let's increase the degree by multiplying two linear ones.

We've already played with

$$(2x - 6)(3x + 6) = 0$$

which, multiplied out, looks like

$$6x^2 - 6x - 36 = 0 .$$

If we want something other than the zeros (roots) of the polynomial, we'd write it in functional form as

$$y(x) = 6x^2 - 6x - 36 .$$

A few things to note here are:

1) This is not a linear equation. For example *y* increases by 36 as *x* goes from 3 to 4, and it increases by 108 as *x* increases from 9 to 10.

2) Since we created this function by multiplying two linear functions, we know by inspection where the zeros of our new function lie. However, second-degree polynomials do not always have two zeros, or the zeros may be imaginary. More on this later.

---

[8] Should $c = 0$ and $a \neq 0$, then $x = {}^{-d}\!/_{a}$ which is the equation of a line parallel to the *y* axis, and is, by definition, not a function. If *a* and *d* should equal zero, we would only have a line on the *x*-axis—i.e., *y* = 0. If *a* = 0 and *d* is finite then we have a horizontal line at $y = {}^{-d}\!/_{c}$ . Obviously, if *a* and *c* should equal zero then *d* too would have to be zero, and you'd have the interesting equation $0 = 0$.

[9] Equations (problems) with more than one variable.

[10] Second-degree equations are often called **quadratic equations**. Quadratic derives from the Middle English word, quadrate, which means "square."

3) If we had only had the equation $6x^2 - 6x - 36 = 0$, we could easily have canceled a 6 out of it and not changed the zeros of the polynomial, but we would have drastically changed the value of $y$ for any $x$ that is not a root of the equation.

Our note #1 is obvious and we need say no more about it.

Note #2, however, deserves more discussion.

A reasonably straightforward equation like

$6x^2 - 6x - 36 = 0$

can be easily factored. You'd first factor out a six to get

$6(x^2 - x - 6) = 0$

which is the same as

$x^2 - x - 6 = 0$.

Since the squared term has *one* as its coefficient, then the $x$ terms in parenthesis also have *ones* as coefficients. Of course, they could be minus *ones*, which you could change by multiplying through by minus *one* two times.

All this considered, you have

$(x + a)(x + b) = 0$.

All you must do is determine what values of $a$ and $b$, added and multiplied, give (for this particular equation)

$a + b = -1$

and

$ab = -6$.

That's not too hard.

Some other easy examples come in the form

$a^2x^2 - b^2 = 0$, which easily factors into

$(ax + b)(ax - b) = 0$.

Another easy one is

$a^2x^2 + 2ab + b^2 = 0$

which factors into $(ax + b)^2$. Please note that this equation has only one zero and that is at $x = {}^{-b}\!/_a$.

Check these out for yourself.

But, if you wanted to factor

$$4x^2 - 7x + 4 = 0$$

your life would not be easy unless you digest the next few paragraphs and learn how to **complete the square**.

Let's take a general second-degree equation

$$ax^2 + bx + c = 0,$$

and see if there is a general way to factor it.

We've already seen how easy it is to factor

$A^2x^2 + 2AB + B^2 = 0$ into $(Ax + B)^2$. (We're using uppercase $A$ and $B$ to eliminate some confusion.)

Let's try to put the general second-degree equation into that form.

Take

$$ax^2 + bx + c = 0.$$

Divide through by $a$, and subtract the constant from both sides to get

$$x^2 + \frac{b}{a}x = -\frac{c}{a}.$$

That wasn't very hard, but neither is it useful by itself, so now add $\dfrac{b^2}{4a^2}$ to both sides and you have

$$x^2 + \frac{b}{a}x + \frac{b^2}{4a^2} = -\frac{c}{a} + \frac{b^2}{4a^2}.$$

After determining a common denominator for the righthand side, this equation turns into

$$x^2 + \frac{b}{a}x + \frac{b^2}{4a^2} = \frac{b^2 - 4ac}{4a^2}.$$

Or, to write it a bit differently,

$$x^2 + \frac{b}{a}x + \left(\frac{b}{2a}\right)^2 = \left(\frac{b^2 - 4ac}{4a^2}\right).$$

Comparing this equation to $A^2x^2 + 2AB + B^2y^2 = 0$ (that we saw earlier), we see that we have completed the square. That is,

$$\left(x + \frac{b}{2a}\right)^2 = \left(\frac{b^2 - 4ac}{4a^2}\right).$$

Now take the square root of both sides as

$$x + \frac{b}{2a} = \sqrt{\frac{b^2 - 4ac}{4a^2}} \ .$$

This is only half of it, however, because the square root may be positive or negative. Hence

$$x + \frac{b}{2a} = -\sqrt{\frac{b^2 - 4ac}{4a^2}}$$

is also a valid solution.

By placing all constants on the righthand side of the equation, you have the zeros of the general second-degree polynomial equation. And of course, that means you have factored the equation into its roots

$$root_1 = \frac{-b + \sqrt{b^2 - 4ac}}{2a}$$                    **Remember this one!**  Eq. 3-3

and

$$root_2 = \frac{-b - \sqrt{b^2 - 4ac}}{2a} \ .$$                    **Remember this one!**  Eq. 3-4

These formulas for the roots (hence, the solutions) are called the *Quadratic Formulas*.

Looks easy enough and it is, but before you go off to lunch, look at the stuff under the radical. This stuff under the radical is called the **discriminant** of the quadratic polynomial, and discriminate it does because it tells you if you have more than one value for the roots and even if the roots are real.

Let's look at that discriminant

$$(b^2 - 4ac)$$

and examine three "what ifs."

1)  What if

   $$b^2 = 4ac\,?$$

   No problem. The discriminant is zero and the two roots are both equal to $\frac{-b}{2a}$.

2)  What if

   $$b^2 > 4ac\,?$$

Again, no problem. Just subtract, take the square root, and keep on chugging with the arithmetic.

But

3)  What if

$$b^2 < 4ac\ ?$$

Now, you must take the square root of a negative number. When we tell our spreadsheet to take the square root of a number such as –1, we get an error message. It's a good thing that we're smarter than computers. To make the math work on this one, we have to use that little thing called "*i*," an imaginary number[11], which we discussed in Chapter 1. With the help of our imagination, the roots become

$$root_1 = \frac{-b + i\sqrt{\left|b^2 - 4ac\right|}}{2a}$$

Eq. 3-5

and

$$root_2 = \frac{-b - i\sqrt{\left|b^2 - 4ac\right|}}{2a}.$$

Eq. 3-6

As you may recall, sets of numbers such as

$$r_1 = R + iI$$

along with

$$r_2 = R - iI$$

(where *R* and *I* are real numbers)

are called **complex conjugates** and when multiplied together produce a real number as

$$r_1 r_2 = R^2 + I^2.$$

Try it and see for yourself.

What do these different forms of the discriminant really mean? Let's plot (Figure 3-2) the three *what ifs* as exemplified by the functions below:

What if #1: $x^2 + 4x + 4 = 0$, which when put into the quadratic formulas gives us the single root at $x = -2$.

---

[11] We don't like the term "imaginary," but we're stuck with it. In the quadratic formula, the imaginary numbers are just a way of making the equations work out. In the real world of electrical engineering, imaginary numbers take on a very real meaning when talking about electrical and magnetic waves that are out of phase with each other, but then the electrical engineers use "j" instead of "i".

What if #2: $x^2 + x - 2 = 0$, which when put into the quadratic formulas gives us roots at $x = 1$ and $x = -2$.

What if #3: $x^2 - 2x + 8 = 0$, which when put into the quadratic formulas gives us roots at $x = 1 + i\sqrt{7}$ and $x = 1 - i\sqrt{7}$.

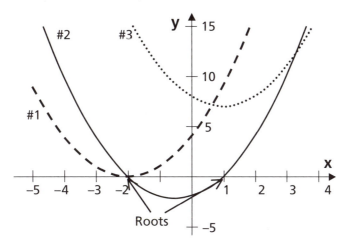

**Figure 3-2. Graph of the quadratic equation for the three cases of the discriminant.**

In *what if #1* the curve only touches the *x*-axis once (at $x = -2$). Where it touches is the single value of the two roots of the equation.

We see that *what if #2* crosses the *x*-axis twice. That's the two roots of the polynomial.

And *what if #3* does not touch the *x*-axis at all. We determined roots for the equation, but those roots do not exist anywhere in real space, hence, they're imaginary.

OK, we've found the roots (zeros, or poles in some technical jargon) of a general quadratic equation. Is that the same as factoring the equation? No, it isn't, but it might be. How's that for an evasive answer?

For a better picture of roots vs. factoring just consider the example that we started with:

$$6x^2 - 6x - 36 = 0$$

which is equal to zero at

$x = 2$ and $-3$,

but these values of *x* are also the roots of the simpler equation

$$x^2 - x - 6 = 0.$$

As you readily see, getting the roots is a good start at factoring, but it's not the whole story.

To get a better grip on this root vs. factoring thing, let's go back to the general form of the quadratic,

$$ax^2 + bx + c = 0,$$

which has roots at

$$x = r_1 \text{ and } r_2.$$

Now look back at the quadratic formulas and note that

$$r_1 + r_2 = {-b}\big/{a}$$

and

$$r_1 r_2 = {c}\big/{a}.$$

If you don't believe us, just plug in the values of $r_1$ and $r_2$ into the sum and product and see for yourself.

With that behind us, we can show that any second-degree (quadratic) polynomial can be written as the product of two first-degree polynomials. Now, this may seem obvious, but let's demonstrate it.

Go back to our general quadratic polynomial,

$$ax^2 + bx + c = y(x),$$

and divide through by $a$ to get

$$x^2 + \frac{b}{a}x + \frac{c}{a} = \frac{y(x)}{a}.$$

Substitute in the values of $\frac{b}{a}$ and $\frac{c}{a}$ in terms of $r_1$ and $r_2$ to get,

$$x^2 - (r_1 + r_2)x + (r_1 r_2) = \frac{y(x)}{a}.$$

Note that this is the same as

$$a(x - r_1)(x - r_2) = y(x).$$

Hence,

$$ax^2 + bx + c = a(x - r_1)(x + r_2).$$

That last part was too easy. Let's look at more interesting stuff. Let's look at some more division of polynomials.

If we want to divide

$$N(x) = ax^n + bx^{n-1} + cx^{n-2} + \cdots + gx + h$$

by

$$M(x) = Ax^m + Bx^{m-1} + Cx^{m-2} + \cdots + Gx + H$$

where

$$n > m ,$$

we'd have a simpler time if $M(x)$ were composed only of factors of $N(x)$.

For example, if $N(x)$ could be factored as

$$N(x) = a(x - e)(x + f)(x + k)$$

and $M(x)$ factored as

$$M(x) = A(x - e)(x + k)$$

then the obvious solution for

$$\frac{N(x)}{M(x)} = Q(x)$$

is

$$Q(x) = \frac{a}{A}(x - f) ,$$

the ratio of the multiplicative constants $(a/A)$ times the only factor of the numerator that was not obliterated by factors in the denominator.

If, however, $M(x)$ has factors not found in $N(x)$, then we have the same situation as in old-time arithmetic division when your denominator has factors not found in the numerator (if you want your quotient in whole numbers, you have a remainder).

Thus, if you divided 7 by 3 your answer is 2 with remainder 1. So in "polynomial speak"

$$N(x) = Q(x) *M(x) + R(x)$$

where $R(x)$ is the remainder.

But let's examine the process. Both $N(x)$ and $M(x)$ are composed of factors of the sort $(x + a)$. The common factors cancel out just as with dividing $(3 * 5 * 7)$ by $(3 * 3 * 7)$ in which case you wind up with $5 / 3 = 1$ with a remainder of 2.

We can really think in terms of dividing the long-winded polynomial, $N(x)$, by only one factor of $M(x)$ at a time.

Let's look back at our general polynomial,

$$N(x) = ax^n + bx^{n-1} + cx^{n-2} + \cdots + gx + h \,,$$

and write it in terms of its factors as,

$$N(x) = a(x + r_1)(x + r_2)(x + r_3) \cdots (x + r_n) \,,$$

where, of course, the $r_i$'s are the roots of the polynomial. How many are they? There are $n$ roots for the $n^{th}$ degree equation. How many different values of $r$ exist? Just as with the quadratic example that we suffered over, at least one and not more than $n$ different values exist. And these values may be real or imaginary.

Stating that the factored form[12] exists is fine, but how do we find these roots? It's not easy unless the degree is 2 or less, you're prescient at divining such things, or you're very lucky. If not, you go to the computer. Although this is not a computer "How To" book, we will touch on numerical analysis and a few computer applications in Chapter 10, Computer Mathematics.

# Examples

## Chemistry

If you are burning ethane (the primary component of natural gas), you will wind up with water vapor and carbon dioxide as the products of combustion. Of course you want to have a sufficient supply of oxygen, so you ask yourself, "How many oxygen molecules do I need for each molecule of methane that I'll burn?" To be safe, you also want to know how much carbon dioxide will be generated.

In such a reaction, the molecules are breaking into atoms or smaller molecules and then recombining as different molecules. Let's look at the starting molecules and the final molecules to see how much oxygen we'll need.

An oxygen molecule has two oxygen atoms, so it's written as $O_2$. Methane has two carbon atoms and six hydrogen atoms, so it's $C_2H_6$. Similarly, carbon dioxide is $CO_2$, and water is $H_2O$. And now for the algebra.

We have

$$C_2H_6 + x\, O_2 = y\, CO_2 + z\, H_2O,$$

where $x$ is the number of oxygen molecules needed to burn each methane molecule, and produce $y$ molecules of carbon dioxide, and $z$ molecules of water vapor. We want to determine $x$, $y$, and $z$.

Here we have one equation and are asked to determine three unknowns. How do we proceed?

---

[12] All polynomials can be factored, although we only showed that factoring works for all first and second degree polynomials.

And the answer is: you proceed very carefully.

Notice that all of the carbon from the methane goes to the carbon dioxide, hence

$C_2$ goes to $y$ C, or

$y = 2$.

Likewise, all of the hydrogen from the methane goes to help make up the water molecules, hence

$H_6$ goes to $z$ $H_2$, or

$z = 3$,

which leaves us with

$C_2H_6 + x\ O_2 = 2\ CO_2 + 3\ H_2O$.

Count up the oxygen atoms on the righthand side of the equation, which equals seven atoms. To get seven oxygen atoms you need $\frac{7}{2}$ oxygen molecules. Most engineers would stop here and state the result as

$C_2H_6 + 7/2\ O_2 = 2\ CO_2 + 3\ H_2O$,

but chemists do not understand fractions and want to write everything in whole numbers. To make them happy, we'll multiply each side by 2 to get

$2\ C_2H_6 + 7\ O_2 = 4\ CO_2 + 6\ H_2O$.

## Economics

You want a ditch around your orchid garden. With much sweat and many blisters you finish the job: it's beautiful and all of your neighbors want one too. You decide to start a ditch-digging business, but wish to accrue no more blisters.

You have a decision to make. You can hire a four-person crew at $25/hour each or buy a $50,000 Ditch-A-Doo ditch-digging machine, which is operated by a $35/hour person, and do the same amount of work as the foursome in the same number of hours. Assuming that enough of your neighbors want ditches that you see work coming your way indefinitely, how many years will it take to break even on the Ditch-A-Doo purchase?

Of course you cannot answer the question until you get more information.

How many hours are worked? We'll assume that the weather is good and the folks work 40 hours for 50 weeks per year.

What, in addition to the operator, does it cost to operate a Ditch-A-Doo? The manufacturer claims that it costs $10 per hour (including saving money to eventually buy a new one). That sounds too low for your conservative nature so you add another $5 per hour to the operating estimate.

What is the interest rate of the loan you'll get to pay for the Ditch-A-Doo? We'll ignore this question (even though it's an important one), and let you ponder interest rates and cost-of-money when you study engineering economics.

To determine the break-even time of your two different approaches you need to write an equation for the costs of each approach as a function of time. The break-even time occurs when these equations equal each other.

So now you can write your algebra equation as:

Costs using ditch diggers is $4 \times \$25H$. Cost using the Ditch-A-Doo is $(\$35 + \$15)H + \$50,000$,

where $H$ is the number of hours worked.

Equating these two we have

$$4 \times \$25H = (\$35 + \$15)H + \$50,000$$

This one is easy. Our first reduction of terms gives us

$$100H = 50H + 50,000$$

where we dropped[13] the dollar signs, and we simplify again by subtracting $50H$ from both sides leaving us with

$$50H = 50,000$$

or

$$H = 1000.$$

Since we assumed in question #1 that 1000 hours is one half a work-year, buying the Ditch-A-Doo is a good idea.

That example seemed too easy, so let's add a new twist to it. Let's throw in a question of ethics. Let's assume that these gardens are in an impoverished country where people are begging for ditch-digging jobs and that the Ditch-A-Doo is built in a highly developed country with nearly zero unemployment. Is it still a good idea to get the Ditch-A-Doo? We won't answer that one.

## Hydrology

(Since hydrology is a civil engineering topic and American civil engineers typically use the English system for units, so shall we in this example.)

---

[13] Actually, we didn't drop the dollar signs. We canceled them out of the equation. Every term had a dollar sign connected to it, so we divided through by $. What we were less careful about was the *per hour* units. The labor costs are stated in $/hour which when multiplied by $H$, the number of hours, gave just $, which are the units as the cost of the Ditch-A-Doo.

Rain is typically measured in inches per hour and land area is measured in acres. The volume of a pond or reservoir is measured in *acre feet*, a volume that is one acre in area and one foot deep.

### Problem 1

Suppose we get a really big rainfall of 5 inches per hour for 4 hours over a 100-acre drainage area. What is the volume that would flow into a proposed reservoir if all of the rain flowed[14] off of the land and into the lake?

*Answer:*

The answers must be in common units so we must change inches of rain to feet of rain as

$$V = 5\frac{in}{hr} * \frac{ft}{12in} * 4hr * 100acres$$
$$= 1600 \;\; acre\text{-}ft$$

Which is to say that if you had a 200-acre reservoir, the water would rise 8 feet.

### Problem 2

Rain falls over a 100-acre watershed with the approximate time distribution shown in Figure 3-3. What is the volume that would flow into a proposed reservoir if all of the rain eventually went into the lake?

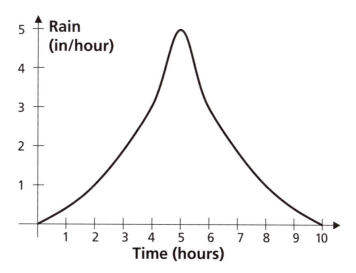

**Figure 3-3. Rainfall vs. Time.**

---

[14] And that's why we chose such a big rainfall. A small rainfall may be largely absorbed into the ground and intervening puddles with little water flowing into the lake. With such a heavy downpour, nearly all of the rain goes into the reservoir.

*Answer:*

We must first calculate the volume of rainfall. This can be done by looking at the area under the curve as such. We can essentially count the squares under the curve.

Amount of rain = 0.5 in/hr × 2 hr + 2 in/hr × 2 hr + 4 in/hr × 2 hr + 2 in/hr × 2 hr + .5 in/hr × 2 hr = 18 inches

Amount = 18 in / 12 in/ft = 1.5 feet

Volume of water running off into the reservoir: 1.5 ft × 100 acres = 150 acre-feet of water.

Additional solved algebra problems relating to engineering can be found on the accompanying CD-ROM.

# *Matrices*

*I'm sorry to say that the subject that I most disliked was mathematics. I have thought about it. I think the reason was that mathematics leaves no room for argument. If you make a mistake, that was all there was to it.* — **Malcolm X**

In the section on polynomials, we spent most of our time worrying how to determine when our function had no value; that is, we looked for all the values of the independent variable that made the function equal to zero. We may have found many values of the independent variable (but, of course, not more than the degree of the equation) but it was always only the one variable that we had to contend with.

Now, let's increase the number of variables. Let's look at two examples[1] and go on from there.

Example 1. The Joe & Jane Shop makes Hawaiian leis. Their small lei has six orchids and sells for $2. The larger one has eighteen orchids and sells for $4. One week they forgot how many leis they sold. They knew, however, that they used 3300 orchids and took in $800. Can you calculate how many leis of each type they made that week?

Example 2. You have made your fortune in the engineering world and are now enjoying your ranch overlooking Waikiki Beach. You decide to devote 576 m$^2$ of your ranch to growing orchids for your friends Joe and Jane, but you only have 120 meters of fencing material. You hate waste, so you decide to use all of the fencing material to make your rectangular plot. What are the dimensions of your orchid garden?

In each of these examples, you have two unknowns. If you only knew how many orchids were used in #1, you would find that many combinations of small and large leis would use 3300 orchids. Likewise, in example #2, if you only knew the area to be fenced, there would be a boundless number of rectangles with an area of 576 m$^2$. Fortunately, you have additional information. Let's take a closer look at these examples and see what the additional information can do for us.

---

[1] And why are we discussing leis in an engineering book, you might ask? We'll present engineering examples later in this chapter. For now let's stay with examples easily visualized to learn the mathematical concepts, and then we'll get more interesting and useful.

Example 1. **Graphing**

Let's write equations for what we know[2].

We'll let $S$ represent the number of small leis and $L$ represent the number of large ones. So,

$6S + 18L = 3300$, the number of flowers used.

If this were all we knew, we could not be of much help. Table 4-1 shows a few of many, many combinations of small and large leis that require 3300 flowers.

**Table 4-1. Combinations of orchids and leis.**

| Small Leis | # of flowers | Large Leis | # of flowers | Total # flowers |
|---|---|---|---|---|
| 370 | 2220 | 60 | 1080 | 3300 |
| 340 | 2040 | 70 | 1260 | 3300 |
| 310 | 1860 | 80 | 1440 | 3300 |
| 265 | 1590 | 95 | 1710 | 3300 |
| 250 | 1500 | 100 | 1800 | 3300 |
| 205 | 1230 | 115 | 2070 | 3300 |
| 130 | 780 | 140 | 2520 | 3300 |
| 100 | 600 | 150 | 2700 | 3300 |
| 70 | 420 | 160 | 2880 | 3300 |

So let's see how the second bit of information can help.

We were told that

$\$2 \cdot S + \$4 \cdot L = \$800$.

Again, many combinations of values for $S$ multiplied by $2 and $L$ multiplied by $4 will yield $800. However, you know that these linear equations have only one set of values for $S$ and $L$ that will satisfy both equations at the same time.

You can solve Joe's and Jane's problem many different ways. You'll first solve it by graphing. Later, you'll manipulate the variables to get the answer. Toward the end of this chapter you'll get elegant and use matrix math to set their books straight.

To make your graph, write each of the known functions in terms of an independent and dependent variable. (If you independently choose a given value for $L$, the value for $S$ is fixed by—that is, dependent on—the equation. Of course, in these examples $S$ could just as well be the independent variable. Just choose the same one for both equations.) So,

$S_1 = -3L_1 + 550$, from the number of flowers equation

---

[2] We believe that this is good advice for approaching any problem, math or otherwise. List what you know and then ask yourself what else do you need to determine how to solve your problem.

and

$S_2 = -2L_2 + 400$, from the money equation. We dropped the dollar symbols to save ink.

You want to determine[3] when $S_1$ equals $S_2$.

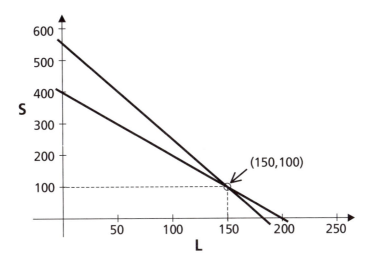

**Figure 4-1. Graph of two linear equations in the search for common values.**

Since these are linear equations, you know that you need to calculate only two sets of points for each line. Calculate those points, connect them, and see where the lines cross as shown in Figure 4-1.

For example, you might have chosen $L = 80$, and then calculated $S$ in the equations for that value of $L$, and marked the graph. Quite likely the value you randomly chose for $L$ did not produce the same value for $S$, the number of small leis, for the two different ways of calculating it. These different answers may well be due to errors in your arithmetic, but more likely, they're due to the nature of coupled equations. Each solution is valid for each equation alone, but for the coupled linear equations, only one solution is valid for both equations.

You then chose another possible value for $L$, say $L = 180$, and calculated two more values for $S_1$ and $S_2$. You plotted these and connected the points. You notice that the lines cross at $L = 150$ and $S = 100$. These are the values of $L$ and $S$ that satisfy both equations. Joe and Jane are happy because they can now show complete books to the IRS, and you are happy because they gave you a double lei that matches the color of your hair!

---

[3]  Since we are not cutting up our flowers, we are dealing only with integers in this graph. However, we'll draw the graph with a continuous line rather than as a sequence of dots.

At this point you might ask, "Can we plot any two related linear equations and determine a pair of values that solve both of them?" Unfortunately, the answer is, "No." Recall that the two lei equations had different slopes, therefore somewhere in the space of orchids, leis, and money the lines had to cross and provide a single solution that satisfied both equations. If, however, the large leis cost $6 and the small ones still cost $2 (that is, the cost was in the same proportion as the number of flowers used in each lei) and they took in $1100, then our two equations would have been

$$6S_1 + 18L_1 = 3300 \text{ (the number of flowers)}$$

and

$$2S_1 + 6L_1 = \$1100 \text{ (the cost of the leis)}$$

When you reduce these equations to the graphing form you have

$$S_1 = -3L_1 + 550 \text{ (for the number of flowers)}$$

and

$$S_2 = -3L_2 + 550 \text{ (for the cost of the leis)}$$

These two different equations, one generated from the number of flowers used and the other made from the revenue brought in, actually make the same line when plotted. Or stated another way, any set of $S$ and $L$ that satisfies the first equation also satisfies the second equation. The second equation is just the linear equivalent of the first. These two equations are said to be **dependent** equations. Your math is correct. It's just that Joe and Jane had the misfortune to price their items in the same ratio that they consumed them. The second equation gave no additional information to help you solve their problem.

Let's consider one more scenario. Suppose Joe and Jane changed the cost information that they gave you and said that the small lei cost $2, the large one cost $6, as in the last example, but that they took in $1000. That is

$$6S_1 + 18L_1 = 3300 \text{ (the number of flowers, as before)}$$

and

$$2S_2 + 6L_2 = \$1000 \text{ (the income from the leis, a new number)}$$

or, simplified,

$$S_1 = -3L_1 + 550 \text{ (for the number of flowers)}$$

and

$$S_2 = -3L_2 + 500 \text{ (for the cost of the leis).}$$

You're so sharp now that you can just look at these last two straight line equations and see that they have the same slope, –3, but have different $y$-axis intercepts ($S$-axis in this example). Since these two lines have the same slope but intercept the vertical axis at different points, they are parallel lines. Although looking down a highway, it may seem like the parallel edges run together near the horizon, you can't find that intersecting point any more than you can find the pot of gold at the end of the rainbow. So what gives here?

It's simple. These two equations do not describe the same physical situation. There is no set of $L$ and $S$ that satisfies both equations. This is an **inconsistent system**. For whatever combination of $L_1$ and $S_1$ we choose, we calculate an income of $1100 for the number of orchids consumed, which is quite inconsistent with the second equation.

Now, let's move on to fencing the orchid garden before we look at other computational techniques.

Example 2.

To recap, you want to place a rectangular fence around a 576 m$^2$ field using 120 m of fencing material. You can write the two related equations for this project, too.

The area of the rectangular field is the product of the depth and width. That is

$$D \cdot W = 576 \, \text{m}^2 \quad \text{(for the area of the plot)}$$

and the length of fencing is just twice the depth plus twice the width, or in math language:

$$2D + 2W = 120 \, \text{m} \quad \text{(for the perimeter of the plot)}.$$

A note on units: You see that one equation has m$^2$ (meters squared) and the other one has units of m (meters). If you are not familiar with keeping track of the units, please learn it now. Tracking the units won't insure the right answer, but ignoring units is a sure path to wrong answers. The side of your rectangle is a length. A length is a measure of the distance from one place to another. Depending on what you are measuring, you may express that length in meters, miles, or even light-years (just to confuse matters, a light-year is the distance that light travels in a year; it is not a unit of time). Now, when we calculate the area of your garden, we multiply two lengths together. The key point with units is that you not only multiply the numbers, but you also multiply the units. This case is simple enough. You multiply the two lengths, each one given in meters, and the result is in m$^2$ or meters squared. (Some use the term *square meters*. We don't like it, but it's the same thing. However, do not ever refer to the area of a plot that is 10 meters on a side as having an area of 10 square meters – it ain't.)

If, however, you were given one of the rectangle's dimensions in meters and the other dimension in feet, how would you multiply meters by feet? It's easy if you understand units. You just convert the feet dimension to meters. (And yes, you could have converted meters to feet, but we're doing metric here.) You look in your table of conversions and see that an inch is no longer defined as 1/12th of the length of the king's foot but is now defined as 2.54 centimeters. A centimeter is defined as 1/100th of a meter, and a meter is defined as the distance traveled by light in a vacuum in 1/299,792,458 of a second. (So using light-year as an example in the last paragraph wasn't too outlandish—pun intended—after all.)

To convert your dimension given in feet to meters, you just step through the conversions.

If the length was $D$ feet then

$$D \text{ feet} \times 12 \frac{\text{inches}}{\text{foot}} = 12D \text{ inches}$$

because the unit, feet, in the numerator cancels with the unit, foot, in the denominator.

Now try for centimeters:

$$12D \text{ inches} \times 2.54 \frac{\text{cm}}{\text{inch}} = 30.48D \text{ cm}$$

because now the inch units cancel.

And now we go to meters.

$$30.48D \text{ cm} \times \frac{\text{meter}}{100 \text{ cm}} = 0.3084D \text{ meters}.$$

We could keep going and express $D$ in light-years if we so chose, but we'll stop here.

Needless to say, we didn't have to take three separate steps to make the conversion. A single chain step would look like

$$D \text{ feet} \times 12 \frac{\text{inches}}{\text{ft}} \times 2.53 \frac{\text{cm}}{\text{inch}} \times \frac{\text{meter}}{100 \text{ cm}} = 0.3084D \text{ meters}.$$

The usefulness of units is by no means limited to length and area. As we plow through this book, we'll reintroduce units whenever we can. Bear with us.

Let's plot these two equations (Figure 4-2) as we did in Example 1. This graph is different from the one for the leis. The equation for the perimeter is linear and graphs as a straight line, but that's not so for our area calculation and graph. That equation is an inverse function that intersects our straight line in two places. Does that mean we have two solutions for this set of coupled equations? Yes, it does.

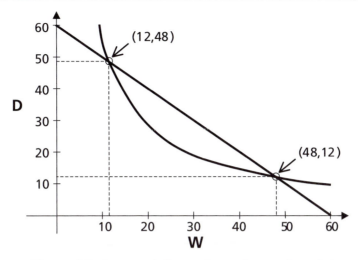

**Figure 4-2. Area and dimensions of a garden plot.**

Which set of *W*s and *D*s is correct? They're both right. The linear equations with their straight lines are restricted to only one solution, but nonlinear equations can have many solutions depending upon the type of equation. Of course, on your Hawaiian ranch, you might not agree to the equivalence of these answers if one of them requires planting orchids in your swimming pool.

As an example of multiple solutions, consider the following two coupled equations[4]

$$y(x) = \sin x$$

and

$$g(x) = \cos x .$$

These two equations are equal whenever

$$\sin x = \cos x .$$

This occurs whenever

$$x = \pi(n + \frac{1}{4})$$

where *n* is any integer, and that creates a lot of answers.

Now we've seen that a set of equations containing at least one nonlinear equation can have more than one solution. You may be tempted to ask if this will always happen. As you might surmise, you may get only one answer or maybe none at all.

---

[4] We've just introduced trigonometry functions, and we cover trig in the next chapter. Skip this example for now if you wish.

If someone had walked off with 24 meters of your fence material, leaving only 96 meters to enclose the 576 m² orchid garden, you would approach your dimension determination the same way. The nonlinear curve would plot just as before. The straight line representing possible rectangular dimensions produced by 96 meters only touches the area curve at one place, and that is at

$$W = D$$

$$= 26 \; meters$$

We'll bet you can guess what would happen if someone walked off with another 10 meters of fencing. That linear curve would fall below the nonlinear curve, and the only solution for enclosing the garden is to put it next to your swimming pool and use part of the pool's fencing to complete the job.

Let's go back to Joe's and Jane's problem and see if there's another way to solve their accounting problem. Recall that they make leis. Their small one contains six orchids and sells for $2, and the larger one contains eighteen orchids and sells for $4. They need to figure out how many leis of each type they sold when all they remember is that they used 3300 orchids and took in $800. Can you calculate how many leis of each type they made that week?

If we call the graphing technique Method 1, then we'll call this one Method 2.

## Method 2. **Substitution**

Given the equations to solve,

$$6S + 18L = 3300 \; \text{(the number of flowers)}$$

and

$$\$2S + \$4L = \$800 \; \text{(the cost of the leis),}$$

solve one equation in terms of one of the variables and insert it into the other equation.

Rewrite the cost of the leis

$$2S + 4L = 800 \; \text{(we've dropped the dollar signs)}$$

as

$$S = -2L + 400$$

The $S$ in the cost equation has the same value as the $S$ in the number of flowers equation, hence we can substitute this $S$ into the first equation as

$$6(-2L + 400) + 18L = 3300 \;.$$

Collecting terms we get

$$6L = 900$$

or

$$L = 150$$

which is the value for $L$ that we determined in the graphing exercise.

Now that we know that Joe and Jane made 150 large leis that fateful week, we can substitute that value for $L$ into either the number of orchids or the revenue equation and determine that they made 100 small leis.

Let's see what happens if we try this method with the dependent equations,

$$6S + 18L = 3300 \text{ (the number of flowers)}$$

and

$$2S + 6L = 1100 \text{ (the cost of the leis).}$$

We'll write the second equation as

$$S = -3L + 550$$

and substitute it for S in the first equation.

$$6(-3L + 550) + 18L = 3300$$

or

$$-18L + 3300 + 18L = 3300,$$

which reduces to $0 = 0$, which is true, but gives us no information

So neither the graphing nor the substitution method works with dependent equations.

"What about the inconsistent equations?" you might wonder. We think you know the answer, but we'll work it anyway:

$$6S + 18L = 3300 \text{ (the number of flowers, as before)}$$

and

$$2S + 6L = 1000 \text{ (the cost of the leis)}$$

So, as before, we write

$$S = -3L + 500,$$

and put it in the top equation to get

$$6(-3L + 500) + 18L = 3300,$$

or

$$-18L + 3000 + 18L = 3300$$

which reduces to

$0 = 300$. Wrong, no good!

To call this result inconsistent is being nice.

> A commentary on Methods 1 and 2. Although Method 2 may seem simpler than Method 1, and it often is, we wanted to start with the graphing technique of problem solving. You've seen this procedure work in finding roots of polynomials. It's visual—you see what's happening—and it's useful in solving other types of problems.

Method 3. **Matrices**

Before entering the world of matrices, let's review a few rules. You've seen parts of these rules used in Method 2, and they're obvious when you think about them.

*Rule #1.* You can interchange equations.

This one is so obvious that it's hardly worth your time. For example, you could have substituted a re-written equation 1 in equation 2 rather than the other way round.

*Rule #2.* You can replace any equation with a non-zero multiple of itself (and don't forget to multiply both sides of the equation).

An example of Rule #2 is just reducing

$2S + 4L = 800$

to

$$\frac{1}{2}(2S + 4L) = \frac{1}{2}(800)$$

which reduces to

$S + 2L = 400$

where we multiplied each side by one-half.

*Rule #3.* You can replace any equation with itself and another equation added to it.

This one deserves a little discussion.

Given the equations for the leis that we've been playing with

$6S + 18L = 3300$ (number of flowers)

and

$2S + 4L = 800$, (cost of the leis)

let's use Rule #2 to multiply the second equation by −3 and then add the result to the first equation. These steps result in

$-3(2S + 4L) + 6S + 18L = -3 \cdot 800 + 3300$

Collecting terms

$$6L = 900$$

or, exercising Rule #2 again and multiplying each side by 1/6,

$$L = 150$$

just as we determined earlier.

Of course, we have fortuitously chosen a couple of steps within the stated rules that simplified our work. Any other addition of one equation to another would have been valid and ultimately led to the solution, but it may not have simplified our lives.

With those rules behind us, let's simplify things. Instead of writing our two equations as

$$6S + 18L = 3300$$

and

$$2S + 4L = 800,$$

let's use the **matrix** notation which is

$$\begin{bmatrix} 6 & 18 \\ 2 & 4 \end{bmatrix} \cdot \begin{bmatrix} S \\ L \end{bmatrix} = \begin{bmatrix} 3300 \\ 800 \end{bmatrix}$$

Nothing too fancy here. We've just written the equations as an array of numbers in the form of three matrices. In this case, the first matrix on the left of the equals sign contains the coefficients of $S$ and $L$, shown in the second matrix. The matrix on the right of the equals sign contains values of the righthand side of the equations. The coefficient matrix is **square** because the number of rows equals the number of columns. The other matrices are **columnar** since all values are in a single column.

Note that coefficients in the top row of the square matrix modify both of the variables in the adjacent columnar matrix, one-on-one, to produce the result shown in the top of the righthand matrix. Or to put it another way, if you take the top row of the first matrix, swing it clockwise ninety degrees, and multiply term-by-term the columnar matrix, you get the value found in the top of the next columnar matrix. This, of course, is where you started. You can then go to the second row, do the same thing, and get the other term in the right-hand columnar matrix. This is matrix multiplication. If we are too wordy here, don't despair. We'll come back to matrix arithmetic in the section on vector calculus.

So, now that you have a shorthand form for writing equations, what are you going to do with it? Matrices won't help you with graphing, but they will aid in the substitution method of solution which you already learned. Here in matrix land we'll call the substitution technique the **Triangular Form**. What we want to do is to use our rules of equation manipulation to have only one non-zero term in the first row, two non-zero terms in the second row, and so forth if you have a larger matrix due to more equations and variables. After we succeed in creating the triangular form, the complete solution is easy.

Let's continue by reducing

$$\begin{bmatrix} 6 & 18 \\ 2 & 4 \end{bmatrix} \cdot \begin{bmatrix} S \\ L \end{bmatrix} = \begin{bmatrix} 3300 \\ 800 \end{bmatrix}$$

to triangular form. We'll use rule #2 to multiply[5] the second row by –4.5, which gives us

$$\begin{bmatrix} 6 & 18 \\ -9 & -18 \end{bmatrix} \cdot \begin{bmatrix} S \\ L \end{bmatrix} = \begin{bmatrix} 3300 \\ -3600 \end{bmatrix}$$

and then we'll use rule #3 to add that product to the top row. Which results in

$$\begin{bmatrix} -3 & 0 \\ 2 & 4 \end{bmatrix} \cdot \begin{bmatrix} S \\ L \end{bmatrix} = \begin{bmatrix} -300 \\ 800 \end{bmatrix}.$$

Reverting to coefficients for a moment, we see that

$$S = 100,$$

which when substituted into row 2 gives

$$200 + 4L = 800$$

or

$$L = 150$$

which are the solutions previously determined. So, what's new?

Using matrices is just a shorthand way of doing the same old stuff. Let's get a bit fancy and systemize the matrix operation.

---

[5] A row in a matrix when multiplied by a constant follows regular multiplication rules. When we multiplied the second row of the first matrix on the left hand side by –4.5, we also multiplied the second row of the columnar matrix on the right hand side by –4.5. We did not multiply both matrices on the lefthand side by the multiplicative factor.

Look at a coupled set of generalized equations with two variables

$$ax + by = t$$

and

$$cx + dy = u$$

where *a, b, c* and *d* are coefficients modifying the variables *x* and *y* to produce *t* and *u*.

Put them into a matrix as

$$\begin{bmatrix} a & b \\ c & d \end{bmatrix} \cdot \begin{bmatrix} x \\ y \end{bmatrix} = \begin{bmatrix} t \\ u \end{bmatrix}.$$

Now let's multiply (rule #2) the first row by *d* to get

$$\begin{bmatrix} ad & bd \\ c & d \end{bmatrix} \cdot \begin{bmatrix} x \\ y \end{bmatrix} = \begin{bmatrix} td \\ u \end{bmatrix}$$

and multiply the second row by *b*:

$$\begin{bmatrix} ad & bd \\ bc & bd \end{bmatrix} * \begin{bmatrix} x \\ y \end{bmatrix} = \begin{bmatrix} td \\ ub \end{bmatrix}$$

and then subtract (rule #3) the second row from the first and restore the original second row which gives

$$\begin{bmatrix} ad - bc & bd - bd \\ c & d \end{bmatrix} * \begin{bmatrix} x \\ y \end{bmatrix} = \begin{bmatrix} td - ub \\ u \end{bmatrix}$$

which reduces to

$$\begin{bmatrix} ad - bc & 0 \\ c & d \end{bmatrix} * \begin{bmatrix} x \\ y \end{bmatrix} = \begin{bmatrix} td - ub \\ u \end{bmatrix}.$$

So here we have the triangular form, and

$$x = \frac{dt - bu}{ad - bc}.$$

Interchange the two rows (rule #1) and follow the same basic procedure and you get

$$y = \frac{au - ct}{ad - bc}.$$

Notice the denominator for both *x* and *y*. It's just a combination of the terms of the square coefficient matrix. Note that the numerator for *x* contains the coefficients of *y* multiplied separately by the right-hand columnar matrix. There is an equivalent pattern for *y*. We'll simplify this even more soon, but to help us along let's introduce the **determinant**.

The **determinant** is a single number that is comprised of the terms you saw in the denominators of both *x* and *y*. It is written as

$$\begin{vmatrix} a & b \\ c & d \end{vmatrix}.$$

As you might have surmised, it is equal to

$$ad - bc.$$

The value of the determinant is calculated by combining the products of the terms in the diagonals. Starting at the top left, you multiply the terms in the diagonal going to the bottom right. You then step over one term and multiply the diagonal in the opposite direction. After the second diagonal multiplication, subtract that product from the first multiplication.

Note the difference in the bracket design to denote a **matrix**, which is an **array** of numbers with definite relationships to variables and equations, from a **determinant**, which is a **single number** derived from a square matrix.

To write the solutions for our generalized *x* and *y* in determinant form, we have

$$x = \frac{\begin{vmatrix} t & b \\ u & d \end{vmatrix}}{\begin{vmatrix} a & b \\ c & d \end{vmatrix}} \quad \text{and} \quad y = \frac{\begin{vmatrix} a & t \\ c & u \end{vmatrix}}{\begin{vmatrix} a & b \\ c & d \end{vmatrix}}$$

Let's go back to our orchid example where we had the equations

$$6S + 18L = 3300$$

and

$$2S + 4L = 800, \text{ which we put in the matrix}$$

$$\begin{bmatrix} 6 & 18 \\ 2 & 4 \end{bmatrix} \cdot \begin{bmatrix} S \\ L \end{bmatrix} = \begin{bmatrix} 3300 \\ 800 \end{bmatrix}.$$

From what we just learned, we can solve the system by evaluating the determinants

$$S = \frac{\begin{vmatrix} 3300 & 18 \\ 800 & 4 \end{vmatrix}}{\begin{vmatrix} 6 & 18 \\ 2 & 4 \end{vmatrix}} \quad \text{and} \quad L = \frac{\begin{vmatrix} 6 & 3300 \\ 2 & 800 \end{vmatrix}}{\begin{vmatrix} 6 & 18 \\ 2 & 4 \end{vmatrix}}.$$

Once more we get the same answers.

This procedure doesn't always work.

Look at the generalized solutions of

$$x = \frac{\begin{vmatrix} t & b \\ u & d \end{vmatrix}}{\begin{vmatrix} a & b \\ c & d \end{vmatrix}} \quad \text{and} \quad y = \frac{\begin{vmatrix} a & t \\ c & u \end{vmatrix}}{\begin{vmatrix} a & b \\ c & d \end{vmatrix}}.$$

If all coefficients were equal to zero, you would be dividing by zero, which is not allowed, but you wouldn't have much of a set of equations to start with. So let's skip that scenario. We'll only consider sets of equations where at least one coefficient of one of the variables is not equal to zero.

The next scenario is to ask what happens if both the denominator and the numerator determinants were equal to zero. Note that the determinants might be zero even though the coefficients are not all equal to zero. Such a condition occurs when we try to solve our dependent example, which produces the following set of determinants:

$$S = \frac{\begin{vmatrix} 3300 & 18 \\ 1100 & 6 \end{vmatrix}}{\begin{vmatrix} 6 & 18 \\ 2 & 6 \end{vmatrix}} \quad \text{and} \quad L = \frac{\begin{vmatrix} 6 & 3300 \\ 2 & 1100 \end{vmatrix}}{\begin{vmatrix} 6 & 18 \\ 2 & 6 \end{vmatrix}}$$

each of which reduces to zero divided by zero. If both denominator and numerator determinants are zero, then you have valid but dependent equations.

What happens if the numerator is non-zero and the denominator is equal to zero? Since you can't divide a non-zero value by zero, you can't determine values for your variables, and we have an inconsistent set of equations. Just for the fun of it, take our inconsistent example (the one where Joe lost money on the horses) and put those coefficients into a matrix. Try to solve for $S$ and $L$ with determinants. It'll be a mess.

We have one more scenario. What do you do with the set of equations shown below?

$$ax + by = 0$$

and

$$cx + dy = 0.$$

This set of equations is similar to our previous set of generalized equations except that these are both equal to zero. These are called ***homogeneous*** equations. Try solving for $x$ and $y$ using determinants. You readily get both $x$ and $y$ equal to zero which is seen by inspection of the equations. These are the only solutions since these are two linear equations, unless they really aren't separate equations. Do we sound as though we speak with forked tongue? Not so. If these two equations are dependent— that is, if the determinant of coefficients is also equal to zero—we really have only one independent equation and

$$x = -\frac{b}{a} y,$$

(or $x = -\frac{c}{d} y$, the same thing in this dependent case)
is a solution.

## More than Two Variables

So far, we have only worked with equations with two variables. Could we have helped Joe and Jane with their pending encounter with the IRS had their leis come in three sizes? Maybe. They would have needed another equation relating the number of leis made to some known value. Maybe we could have helped them if they knew the number of ribbons used on each lei and total number of ribbons they used.

In simpler words, if we have three unknowns (variables) in the equations, we need three independent equations. If we have four unknowns, we need four equations, and so on.

All of the procedures shown before work with three or more unknowns, although graphing gets a bit complicated. We'll only discuss the matrix method as it, in its elegance[6], is the best method to keep track of multiple unknowns. The determinants for solving three unknowns with three equations is the same as when you had only two, except working out the value of the determinant is a bit messier.

As before, we solve for the first unknown by exchanging the first column of the numerator determinant with the righthand columnar matrix and then dividing by the coefficient determinant. The second variable is determined by exchanging that same columnar matrix with the second column and so forth. We are not going through the messy proof as with the two unknowns set, but hopefully this leap to three or more is

---

[6] And that's the value of the matrix notation. It's just a handy bookkeeping procedure.

acceptable to you. If not, work them out and prove it to yourself. We will, however, walk through the procedure for determining your solution.

Starting with three generalized, independent equations of

$$ax + by + cz = t,$$
$$dx + ey + fz = u,$$

and

$$gx + hy + kz = v,$$

we write the matrix form as

$$\begin{bmatrix} a & b & c \\ d & e & f \\ g & h & k \end{bmatrix} \cdot \begin{bmatrix} x \\ y \\ z \end{bmatrix} = \begin{bmatrix} t \\ u \\ v \end{bmatrix}$$

and then solve for $x$, $y$, and $z$ using determinants.

$$x = \frac{\begin{vmatrix} t & b & c \\ u & e & f \\ v & h & k \end{vmatrix}}{\begin{vmatrix} a & b & c \\ d & e & f \\ g & h & k \end{vmatrix}}, \quad y = \frac{\begin{vmatrix} a & t & c \\ d & u & f \\ g & v & k \end{vmatrix}}{\begin{vmatrix} a & b & c \\ d & e & f \\ g & h & k \end{vmatrix}}, \quad \text{and} \quad z = \frac{\begin{vmatrix} a & b & t \\ d & e & u \\ g & h & v \end{vmatrix}}{\begin{vmatrix} a & b & c \\ d & e & f \\ g & h & k \end{vmatrix}}.$$

You might question how to determine the value of one of these larger determinants. It's really the same procedure as before:

$$\begin{vmatrix} a & b & c \\ d & e & f \\ g & h & k \end{vmatrix} = a \cdot \begin{vmatrix} e & f \\ h & k \end{vmatrix} - b \cdot \begin{vmatrix} d & f \\ g & k \end{vmatrix} + c \cdot \begin{vmatrix} d & e \\ g & h \end{vmatrix}.$$

We determine the value of the determinant of a 3×3 matrix by placing imaginary lines, one horizontal and one vertical, across each term in the top row and then multiplying that value times the value of the 2×2 determinant not covered. As you can see, we alternate the sign as we step across the top row. You could have chosen any row or column to step across in the determination of the value of the determinant.

Should you be unfortunate enough to have to solve for four or more unknowns without the aid of a math program on a computer, you solve for the value of the determinant just as you did in the 3×3 case. You draw your imaginary lines through the top left number and pretend that you have a 3×3 determinant to solve. You solve it and multiply that value by the value that you had covered. Then you move your imaginary lines to the next value in the top row (or you could move down the column—

it doesn't matter which direction you go as long as you are consistent), change its sign, and repeat until you've marched across the top row.

## Some Matrix Arithmetic

We've used matrices and evaluated determinants of matrices. Now let's take a little break and look at some of the algebra of matrices.

First, some definitions and rules:

- A matrix is an ordered set of values.

- If it has $m$ rows (a row is one of the horizontal lines of numbers) and $n$ columns (vertical lines) the matrix is an $m \times n$ matrix.

- If $m = n$, we have a square matrix.

- Determinants can come only from square matrices.

- The quantities within the matrix are called the elements of the matrix.

- Two matrices are identical if, and only if, each of the elements are identical. That is, if matrix $A$ is

$$A = \begin{bmatrix} a_{11} & a_{12} & a_{13} \\ a_{21} & a_{22} & a_{23} \end{bmatrix}$$

and matrix $B$ is

$$B = \begin{bmatrix} b_{11} & b_{12} & b_{13} \\ b_{21} & b_{22} & b_{23} \end{bmatrix}$$

then $A = B$ if and only if $a_{11} = b_{11}, a_{12} = b_{12}, a_{22} = b_{22}, \cdots, a_{mn} = b_{mn}$.

- Unequal matrices may have the same determinant, and of course, equal matrices will have the same determinant.

- You can add or subtract matrices. They must, however be the same size. To add $A + B$, you make a new matrix $C$ where each element of $C$ is the sum of the same elements in $A$ and $B$. That is

$$A + B = C$$
$$= \begin{bmatrix} a_{11} & a_{12} & a_{13} \\ a_{21} & a_{22} & a_{23} \end{bmatrix} + \begin{bmatrix} b_{11} & b_{12} & b_{13} \\ b_{21} & b_{22} & b_{23} \end{bmatrix},$$
$$= \begin{bmatrix} a_{11} + b_{11} & a_{12} + b_{12} & a_{13} + b_{13} \\ a_{21} + b_{21} & a_{22} + b_{22} & a_{23} + b_{23} \end{bmatrix}$$

and it follows that

$$A + B = B + A.$$

■ You can multiply matrices. You just don't want to do so unless you can stick them into a computer math program. There are some rules. For example, you cannot multiply $A$ and $B$ as seen above. You can only multiply a $m \times n$ matrix with a $n \times p$ matrix, and the product is a $n \times p$ matrix (where $p$ need not equal $m$ or $n$). So let's get a new matrix $D$ of size $2 \times 2$ and multiply it times $A$. That is

$$DA = E$$

$$= \begin{bmatrix} d_{11} & d_{12} \\ d_{21} & d_{22} \end{bmatrix} \cdot \begin{bmatrix} a_{11} & a_{12} & a_{13} \\ a_{21} & a_{22} & a_{23} \end{bmatrix} \qquad \textbf{Remember this one!}$$

$$= \begin{bmatrix} d_{11}a_{11} + d_{12}a_{21} & d_{11}a_{12} + d_{12}a_{22} & d_{11}a_{13} + d_{12}a_{23} \\ d_{21}a_{11} + d_{22}a_{21} & d_{21}a_{12} + d_{22}a_{22} & d_{21}a_{13} + d_{22}a_{23} \end{bmatrix}$$

Hence, unless the matrices are square, you cannot reverse the order of multiplication, and even if they are square, they, in general, are not commutative. That is, in most cases $AE \neq EA$.

■ A diagonal matrix is one where all elements are zero except on the main diagonal, such as

$$D = \begin{bmatrix} d_{11} & 0 \\ 0 & d_{22} \end{bmatrix}.$$

■ You would have a scalar matrix if $d_{11} = d_{22}$ in the diagonal matrix. Note that multiplying a scalar matrix times another matrix of proper size will increase the value of each element of that matrix by the value of the element of the scalar matrix. Such as

$$DA = E$$

$$= \begin{bmatrix} d & 0 \\ 0 & d \end{bmatrix} \cdot \begin{bmatrix} a_{11} & a_{12} & a_{13} \\ a_{21} & a_{22} & a_{23} \end{bmatrix}$$

$$= \begin{bmatrix} da_{11} & da_{12} & da_{13} \\ da_{21} & da_{22} & da_{23} \end{bmatrix}$$

■ If the elements in a diagonal matrix all equal one, you would have an identity matrix. That is, multiplying a matrix by an identity matrix does not change its value.

■ If you interchange rows and columns in a matrix, you have the transpose of that matrix. That is, if

$$A = \begin{bmatrix} a_{11} & a_{12} \\ a_{21} & a_{22} \end{bmatrix}$$

then the transpose of *A* (which we call A′) is

$$A' = \begin{bmatrix} a_{1,1} & a_{2,1} \\ a_{1,2} & a_{2,2} \end{bmatrix}.$$

Is all this really used to solve problems?

Yes, it is, and you have used it. You used matrix multiplication when you determined the number of small vs. large leis that Jane and John sold. Do you remember the sets of matrices of the form

$$\begin{bmatrix} c_{11} & c_{12} \\ c_{21} & c_{22} \end{bmatrix} \cdot \begin{bmatrix} x_1 \\ x_2 \end{bmatrix} = \begin{bmatrix} n_1 \\ n_2 \end{bmatrix},$$

where the *C* matrix contained the coefficients for the number of flowers and the costs[7] of each type of lei, the columnar *X* matrix was the unknown number of small and large leis sold, and the *N* matrix represented the total number of flowers and the total dollars?

If you multiply the *X* matrix by the *C* matrix you have

$$\begin{bmatrix} c_{11}x_1 + c_{12}x_2 \\ c_{21}x_1 + c_{22}x_2 \end{bmatrix} = \begin{bmatrix} n_1 \\ n_2 \end{bmatrix},$$

and since these two matrices are equal, the elements must be equal. Hence,

$$c_{11}x_1 + c_{12}x_2 = n_1$$

and

$$c_{21}x_1 + c_{22}x_2 = n_2$$

which are the two equations that you started with.

Also, when you solved for $x_1$ and $x_2$ using matrices and determinants, you not only performed matrix multiplication, but you also inverted a matrix. However, since you know how to solve for those unknowns using matrices, and you saw that doing so was consistent with mathematics, we will not drag you through matrix inversion.

---

[7] "Whoa," you might say. "Are we mixing units? You have dollars on one line and flowers on another."
Good observation. But look again. You have no units. One equation has *dollars plus dollars equals dollars*, and the other equation has *flowers plus flowers equals flowers*. You are consistent with your units in each equation, but once you have a certain unit across the equation, the unit cancels out. In these cases, the *dollars* or *flowers* cancel. You still have implicit units such as *number of small leis* in your equations.

### Example: Electrical Engineering

We'll start where all electrical engineers start: circuits with batteries and resistors as shown in Figure 4-3.

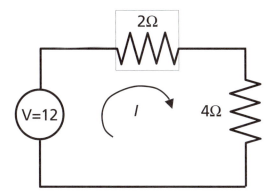

**Figure 4-3. A simple electrical circuit.**

As you see, the circuit has a 12-volt battery and a 2-ohm resistor in series with a 4-ohm resistor. The controlling equation for this circuit comes from Ohm's law, which states

$$V = I \cdot R .$$

where

$V$ = voltage (volts) (see footnote[8])

$I$ = electrical current (amps)

and

$R$ = resistance (ohms)

Since we know the voltage and total resistance[9] we can calculate[10] the current in the circuit and the voltage drop across each resistor. We start with the current which is

---

[8] Voltage is often referred to as $E$ or $e$ in engineering books. The $E$ comes from electromotive force or *emf*.

[9] The resistance of resistors add when they are connected in series as shown in Figure 4-3. Hence, the total resistance of this circuit is 2 + 4 = 6 ohms.

[10] With our limited information, the current is all that we can calculate, but that is certainly not all that we might want to know. The total energy content of the battery, the heat dissipation of the power consumed in the resistors, and the hydrogen evolution and other environmental aspects of the battery are also important to good design.

$$I = \frac{V}{R}$$

$$= \frac{12}{(2+4)} \quad .$$

$$= 2 \text{ amps}$$

Now that we know the current, we can calculate the voltage drop across each resistor from $V_1 = I \cdot R_1$ and $V_2 = I \cdot R_2$. We can now step around the circuit and see that the total voltage drop across the two resistors equals the applied battery voltage.

That was a simple one-mesh circuit. Let's do a two-mesh circuit as shown in Figure 4-4.

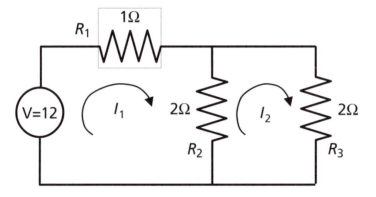

**Figure 4-4. A two-mesh electrical circuit.**

We have the same 12-volt battery, and it has to supply current to two loops of resistors. As we had a clockwise current in the one-mesh circuit, we're assuming two clockwise currents in the two-mesh circuit. We have $I_1$ leaving the battery and going clockwise into loop one and we have $I_2$ circling clockwise in loop #2. The tricky part is designating the current through resistor #2. We have $I_1$ going into the top of that resistor in the clockwise direction. $I_2$ is coming into the bottom of resistor #2, which from the perspective of loop #1 is the counter-clockwise direction. These two currents try to cancel each other. Hence, the net current going through resistor #2 is $I_1 - I_2$.

Now we write the loop equations similar to what we did in the one-mesh case:

$$V = I_1 \cdot (R_1 + R_2) - I_2 \cdot R_2$$

for loop #1, and

$$0 = -I_1 \cdot R_2 + I_2 \cdot (R_2 + R_3),$$

for loop #2. (Loop #2 has no battery, so the voltage drops across the resistors in this loop must sum to zero.)

Now we have two equations, and we can solve for the two unknowns, $I_1$ and $I_2$. We can solve for the two currents by graphing or by substitution, but let's use matrices.

$$\begin{bmatrix} R_1 + R_2 & -R_2 \\ -R_2 & R_2 + R_3 \end{bmatrix} \cdot \begin{bmatrix} I_1 \\ I_2 \end{bmatrix} = \begin{bmatrix} V \\ 0 \end{bmatrix}$$

and we calculate $I_1$ and $I_2$ as

$$I_1 = \frac{\begin{vmatrix} V & -R_2 \\ 0 & R_2 + R_3 \end{vmatrix}}{\begin{vmatrix} R_1 + R_2 & -R_2 \\ -R_2 & R_2 + R_3 \end{vmatrix}}$$

$$= \frac{\begin{vmatrix} 12 & -2 \\ 0 & 4 \end{vmatrix}}{\begin{vmatrix} 3 & -2 \\ -2 & 4 \end{vmatrix}}$$

$$= \quad 6 \text{ amps}$$

and

$$I_2 = \frac{\begin{vmatrix} R_1 + R_2 & V \\ -R_2 & 0 \end{vmatrix}}{\begin{vmatrix} R_1 + R_2 & -R_2 \\ -R_2 & R_2 + R_3 \end{vmatrix}}$$

$$= \frac{\begin{vmatrix} 5 & 12 \\ -2 & 0 \end{vmatrix}}{\begin{vmatrix} 3 & -2 \\ -2 & 4 \end{vmatrix}} \quad .$$

$$= \quad 3 \text{ amps}$$

We can now calculate the voltage drop across any resistor with Ohm's law, $V_i = I_i \cdot R_i$, where the subscript, $i$, refers to the resistor of interest. For example, the voltage drop across $R_2$ is

$$V_2 = (I_i - I_2) \cdot R_2$$
$$= (6 - 3) \cdot 2 \quad .$$
$$= \quad 6 \text{ volts}$$

We've discussed some techniques to solve problems with more than one unknown. We have graphed, substituted, and introduced matrices and determinants to solve these multivariable sets of coupled equations. The beauty of all this is that we always get the same answer regardless which approach we use.

You may wonder why we introduced the matrix concept. The examples we used could have been solved by other means. With the two unknowns, two equations set, you need not bother with matrices, and solving the three unknowns, three equations sets may be a toss-up between solution techniques. However, with anything larger, and especially when applied to computers, the matrix operation is the only way to go.

You may also wonder if there are even more uses for matrices than what we've shown. You bet there's more, as you'll see in the chapters on vector calculus and computer math.

# *Trigonometry*

*The mind that constantly applies itself to the study of lines, forms and angles is not likely to fall into error. In this convenient way, the person acquires intelligence.*
— **Ibn Khaldun, In** *The Muqaddimah. An Introduction to History*, **1375**

Trigonometry is one of the longer words in our math vocabulary, but it has a simple definition: it simply means a triangle measurement. The definition is appropriate since trigonometry is very useful for measuring triangles, and it is also quite useful for describing many periodic occurrences. We can use trigonometry to describe such diverse phenomena as the steepness of a ski slope, properties of ocean waves and characteristics of light transversing the heavens.

We've said that trigonometry means triangle measurement, and we know that tri-angle means three angles, so let's define **angle**.

First, draw a horizontal line and call it the $x$-axis as we did in Chapter 2 on functions. Now, draw the $y$-axis which, of course, is perpendicular to the $x$-axis. You have just made four right angles[1].

Next, swing a line clockwise up from the x-axis, pivoting about the origin of your graph, but don't swing as far as the y-axis. Draw this line through the origin and then an equal distance into the third quadrant. You have now created a bunch more angles.

Obviously, a "bunch" isn't very elegant mathematical terminology. We'll name a few of the angles, but as you'll see later, since these angles can repeat forever, saying "a bunch" isn't too bad.

Let's call the angle between the $x$-axis and our new line, $\theta$, and we'll call the angle between our line and the $y$-axis, $\alpha$. Note that if you add $\alpha$ to $\theta$ you have a right angle. Any two angles that sum to a right angle are called ***complementary angles***; we can say that $\theta$ is *complementary* to $\alpha$ or the other way round. Figure 5-1 helps to sort out the terminology.

---

[1]  Actually, we can say that you made 6, 8, or millions of angles. More on that later in this chapter. We'll stick to four angles for now.

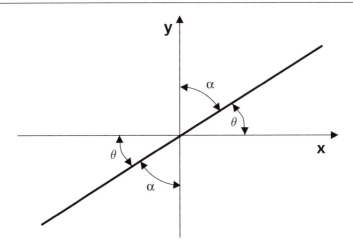

**Figure 5-1. The x-y coordinate system and angles.**

Angle $\theta$ is smaller than a right angle. Such angles are called **acute angles**. Let's name the angle that swings from where $\theta$ stops all the way to the negative *x*-axis, $\beta$. $\beta$ is larger than a right angle and is called an **obtuse angle**. Note that adding $\beta$ to $\theta$ makes a straight line. Any two angles which, when added together, make a straight line are called **supplementary angles**.

Take a look in the lower two quadrants of your graph. Note that angles with the same values as $\theta$, $\alpha$, and $\beta$ also exist down under (Figure 5-2).

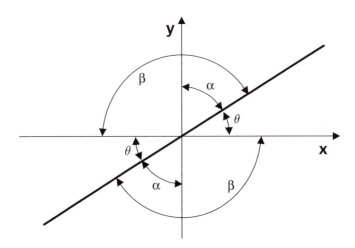

**Figure 5-2. Similarity of angles in different quadrants.**

Now, draw a circle with a radius of five; use whatever units you desire. Assign the center of the circle to have coordinates of 0,0 and draw an *x* and a *y* axis centered on the circle.

You can note that the circle cuts the *x*-axis at (5,0) and (–5,0) as well as the *y*-axis at (0,5) and (0,–5). Now go out four units in the positive *x*-direction and then extend an invisible vertical line in the positive *y*-direction until it intersects your circle. Mark the circle where your invisible line crosses it. Next measure three units vertically on the positive *y*-axis and extend an invisible horizontal line in the positive *x*-direction until it hits your circle[2]. These two intersections of the circle are the same point, which is (4,3). Your next assignment is to draw a visible line from the coordinate (4,3) through the origin (0,0), and, just for fun, continue your line into the third quadrant to intersect the circle on the opposite side (Figure 5-3). Measure the coordinates of that intersection. Hopefully they are (–4,–3).

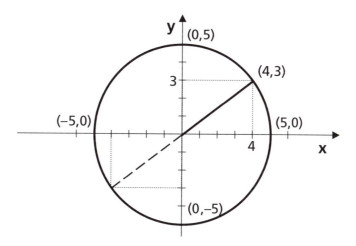

**Figure 5-3. Triangle measurements as parts of a circle.**
**The invisible lines mentioned in the text are shown as dashed lines.**

How does that quantify the angle? You can look at the figure and say that your line has swung or tilted a bit more than 1/10th of the circle away from the positive *x*-axis and the extended line has swung a bit more than 1/10th of the circle past the negative *x*-axis.

---

[2] You don't have to draw your circle around the origin of the graph, but it's much simpler this way. And there's nothing special about the 3,4,5 lengths except that connecting those lengths form a right triangle with easy numbers to work with.

The tilt between these lines is an angle, and we could use fractions like these to quantify the measurements, but there are much better ways. We'll use the conventional measurements of degrees and radians. Degrees are the more common measurement[3] used in lay language, and measurements in degrees are easier to visualize. Radians are usually preferred in engineering and science, and you'll see why shortly. We'll spend a few minutes on degrees and then go on to radians.

If we walk around the circumference of our circle, starting at (5,0) and divide that length into 360 equal parts, then each time our magic swing line moves from one of the 360 measured tic marks to another, it has moved one degree. The line that we drew from (4,3) through (0,0) should hit the circle just a smidgen below 37° as shown in Figure 5-4.

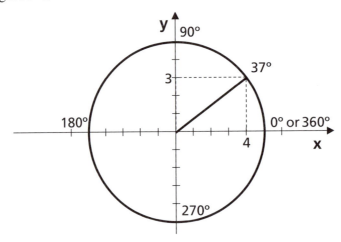

**Figure 5-4. Angles measured in degrees.**

If you want finer measurements for your angles than degrees, you can divide each degree into 60 equal steps called minutes[4]. To get *really* high angular resolution, you can then divide each minute into 60 seconds. So far we've used existing terminology for temperature and then time to designate angles, and now we have more terms. The symbols for minutes and seconds, in angular measurement, are the same as the Americans use for feet and inches. For example, if you wanted to write 60 degrees 20 minutes and 15 seconds using the accepted symbols, it would be written as 60° 20' 15".

---

[3] Note that we say measurement and not units. Regardless of how we measure our angles, we never get away from portions or fractions of the circle. Hence, whatever units the circle may represent, when we divide those units into our portion, in the same units, the units cancel each other. That is, angles are dimensionless.

[4] Just a note: a nautical mile is defined as one minute of longitude at the equator. Hence the circumference of the earth is 60 × 360 = 21,600 nautical miles.

Let's note a couple of things about angle measurement before we move on.

1) Our swinging line doesn't have to move counterclockwise. A swing of 270° counterclockwise describes the same angle as a swing of 90° clockwise. The swing was greater to get to this position, but the angle is defined as the displacement from the x-axis. We will always measure our angles from the positive x-axis in the counterclockwise direction. There's nothing magic about this. It's like driving. The English aren't wrong to drive on the lefthand side of the road in their own country, but when they go to the continent they find life much easier following the local convention.

2) Measurements of angles are just that—measurements between angles—and we do not have to reference them to the positive x-axis.

3) Angles of 360° plus $x°$ are the same as $x°$ measured from the positive x-axis. We'll get to periodic functions later in this chapter, but do you see a hint of what's coming?

4) We may be jumping a bit ahead, but note that our magic line for each angle intersects the circle at an unique coordinate $(x_i, y_i)$. Hence, if we specify the radius of the circle and the coordinate $(x_i, y_i)$ we have uniquely specified the angle (unless, of course, you want to add 360° to it). See Figure 5-5.

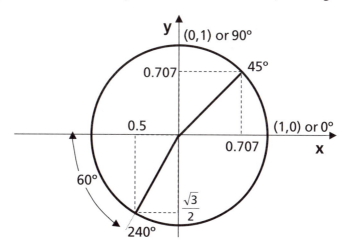

**Figure 5-5. Angle measurements.**

5) Also note that $x_i^2 + y_i^2 = r^2$ for any angle drawn in our circle. This equation was given to us by the Greek mathematician Pythagoras, and is called the Pythagorean theorem. This can be seen by dropping a vertical line from $(x_i, y_i)$ to the x-axis to form a triangle in which one of the angles is 90° (which we call a right triangle). The **Pythagorean theorem** states that the length of the longest side of a right triangle is the square root of the sum of the squares of each of the shorter two sides.

6) Look at Figure 5-6 with the 60° angle in it. We see our magic line at 60° and if we swing our line another 30° we would have a vertical line. Hence for any angle[5], $\theta$, in the first quadrant, a **complementary angle** of 90° − $\theta$ exists between our line and the y-axis. We also have a **supplementary angle** of 180° − $\theta$ that goes from $\theta$ to the negative x-axis.

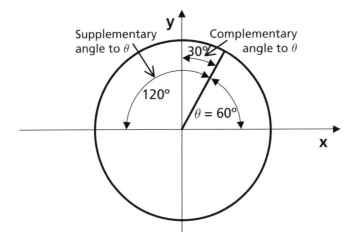

**Figure 5-6. Complementary and supplementary angles.**

7) Extend your magic line to the opposite side of the circle. Note that the same angles, as measured from the negative *x*-axis, exist in the lower two quadrants.

8) Should our magic line describing some angle less than 90° extend through the origin of the graph and go on to intersect the circle in the third quadrant, the coordinates of intersection would be the same as the intersection in the first quadrant except for the signs.

And that's enough said about degrees. What about radians?

## Radians

As we suspect you know, the circumference of a circle divided by its radius is equal to $2\pi$[6]. We are going to divide the circumference of our circle into $2\pi$ parts and say that we have $2\pi$ radians (of angle) in our circle.

---

[5] The Greek letter $\theta$ seems to be the universal symbol for an angle, just as *x* is the symbol for an unknown quantity. We'll of course use other symbols when talking about more than one angle.

[6] $\pi$ is that irrational number approximately equal to 3.14159265358979. We usually drop a few of those decimal places and call it 3.14.

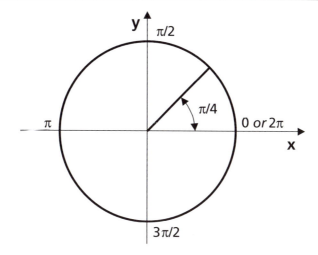

**Figure 5-7. Angles measured in radians.**

Another way to look at the radian is that it represents the length of an arc[7], along the circumference of a circle, that is equal to the circle's radius. If you think that this sounds silly, we'd have to agree with you at this point. However, it'll make more sense later in the chapter.

Simple comparisons of our angle measurements are given in Table 5-1.

**Table 5-1. Angle measurement comparisons.**

| Portion of a Circle | Degrees | Radians |
|---|---|---|
| 1/6 | 60 | $2\pi/6$ |
| 1/4 | 90 | $\pi/2$ |
| 1 | 360 | $2\pi$ |

## Some Special Relationships

Remember our brief discussion in which a line intersecting a circle has a unique set of coordinates $(x_i, y_i)$ for each angle within the first[8] $2\pi$ of angle creation? Hence, if we specify the radius of the circle and the coordinate $(x_i, y_i)$ we have uniquely speci-

---

[7] Don't confuse arc with chord. Both are lengths. The arc is measured along the circumference of the circle. It's curved. The chord is the straight line distance between two points on a circle.

[8] You can go around the circle as many times as you want. If you start at $\pi/4$ and go around the circle back to your starting point, you have a new angle of $2.25\pi$. Go around again and you have $4.25\pi$. It never stops, but the coordinates on the circle stay the same. We'll stay within the first $2\pi$ radians unless otherwise stated.

fied the angle (unless, of course, you want to add an integral multiple of $2\pi$ radians
to it). Let's take that triangle that we created earlier (Figure 5-3) that had one point at
(0,0), another point at (4,0), and the third one at (4,3). As we discussed earlier, the
vertical line from (4,0) to (4,3) defines a right angle with respect to the *x*-axis, so we
call this triangle a **right triangle**. We'll also give the lines themselves some names.
The line from (0,0) to (4,0) we'll call the **line adjacent** or just the adjacent. The
vertical line from (4,0) to (4,3), the line opposite the origin, we'll call the **line opposite**
or the opposite. The longest line, the one from (0,0) to (4,3), is the **hypotenuse**.
Since we're working with an *x-y* coordinate graph we can shorten these names even
more. We'll call the adjacent *x*, the opposite *y*, and the hypotenuse *r*.

And now for the *really* special names. You can probably forget most of the special
names we've given you so far, but the next three names must make indelible wrinkles
into your brain forever, and the subsequent three should at least make small creases.

The important names are *sine*, *cosine*, and *tangent*, which you'll abbreviate as
sin, cos, and tan. The slightly less important names are secant, cosecant, and cotangent, which are abbreviated as sec, csc, and cot. Now, let's define them.

Since we are talking about angles, we define them for the angle, here named $\theta$,
which is the angle between the *x*-axis and the hypotenuse. We hope that Figure 5-8
helps in creating your wrinkles. For the time being, we are keeping $\theta \le 90°$.

First:

$\sin\theta = {}^{y}\!/_{r}$ , called "sine theta"

$\cos\theta = {}^{x}\!/_{r}$ , called "cosine theta"  *(Remember these.)*

and

$\tan\theta = {}^{y}\!/_{x}$ , called "tangent theta" (or "tan theta").

The lesser names are just manipulations of these three. They are

$\sec\theta = {}^{r}\!/_{y}$ , called "secant theta"

$\csc\theta = {}^{r}\!/_{x}$ , called "cosecant theta"

and

$\cot\theta = {}^{x}\!/_{y}$ , called "cotangent theta" (or "cotan theta").

These relationships are called **trigonometric functions**[9], or trig functions for
short.

---

[9] Values for these trig functions are listed in tables and are readily available from your
computer and many calculators. Should you have the value of one of these functions and
want to know what angle produced it, then you look for the arc(trig function). For
example $\sin 45° = 0.707$, and $\arcsin(0.707) = 45°$.

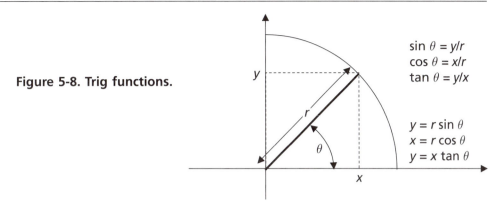

**Figure 5-8. Trig functions.**

$$\sin \theta = y/r$$
$$\cos \theta = x/r$$
$$\tan \theta = y/x$$

$$y = r \sin \theta$$
$$x = r \cos \theta$$
$$y = x \tan \theta$$

The most important thing to note is that the values of these functions are only dependent on $\theta$. You can make the circle any size you want to and the ratios that define these functions do not change.

The next thing to consider is the relationship between $\theta$ and $\alpha$, where $\theta$ is the angle between the *x*-axis and the hypotenuse, and $\alpha$ is the third angle in the right triangle. Note that we are still limiting ourselves to an acute angle for $\theta$.

Remember that an angle equal in magnitude to $\theta$ exists between the negative *x*-axis and the extension of the hypotenuse into the third quadrant, so we have a second right triangle with the same interior angles. Slide that second right angle (see Figure 5-9) along the hypotenuse to form a rectangle as shown. We now have a new fact about right triangles: the two smaller angles are complements of each other. That is, $\theta = \dfrac{\pi}{2} - \alpha$. And from that, it follows that the three interior angles of a right triangle sum to $\pi$ radians[10].

It also follows that in a right triangle

$$\sin\theta = \cos\alpha$$

$$\cos\theta = \sin\alpha$$

and

$$\tan\theta = \cot\alpha$$

along with similar relationships for the secants. (You can work them out if you're interested.)

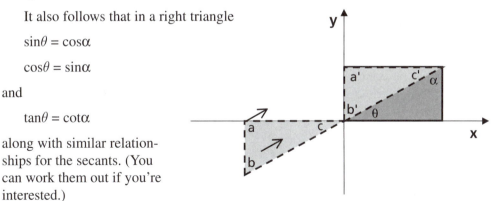

**Figure 5-9. Formation of a rectangle from two right angles.**

---

[10] The interior angles of any triangle sum to $\pi$ radians. You'll show this in a few more pages.

# The Distance Equation

What we do next may seem like another digression, but bear with us. We're going back to Pythagoras and his theorem, but this time we'll use sines and cosines.

We have defined $\sin\theta$ as the ratio of the values of $y$ and $r$, where $y$ is the value of the $y$-axis coordinate where the line defining the angle $\theta$ crosses our circle, and $r$ is the length on the triangle's hypotenuse which, of course, is equal to the radius of the circle. We have also defined $\cos\theta$ in a similar manner.

Consequently, if we know the radius of the circle, or the hypotenuse, as well as the sine and cosine for whatever angle we're working with, we know $(x,y)$ from:

$$x = r\cos\theta$$

and

$$y = r\sin\theta.$$

And here is where we call upon our friend Pythagoras, and he tells us that

$$x^2 + y^2 = r^2.$$

We now see that

$$(r\cos\theta)^2 + (r\sin\theta)^2 = r^2,$$

or simplified

$$\cos^2\theta + \sin^2\theta = 1,$$

where $\cos^2\theta = \cos\theta \times \cos\theta$.

This is an important trigonometric identity. *Remember it.*

Now consider two different angles $\theta$ and $\varphi$ where $\theta$ makes a triangle within our old circle of radius $r$, and $\varphi$ makes a triangle within a circle of radius $A$. Both circles are centered at $(0,0)$ (see Figure 5-10). Now we want to know the distance from where the line describing angle $\theta$ touches its circle, which we'll call $P(\theta)$, to the point where the line describing angle $\varphi$ hits the second circle, which we'll call $P(\varphi)$. We want to express this distance with sines and cosines.

As before, the coordinates, $(x,y)$, of $P(\theta)$ are $(r\cos\theta, r\sin\theta)$, and the coordinates of $P(\varphi)$ are $(A\cos\varphi, A\sin\varphi)$. Now comes the tricky part. Look at the distance between those two points as the hypotenuse of a new triangle as shown in Figure 5-10. The magnitude[11] of the adjacent side ($x$-axis) of this new triangle is

$$|\, r\cos\theta - A\cos\varphi \,|$$

---

[11] We use the term magnitude because in the general case $A\cos\alpha$ might be larger than $r\cos\theta$ which would produce a negative number. That's OK since we are going to square the result.

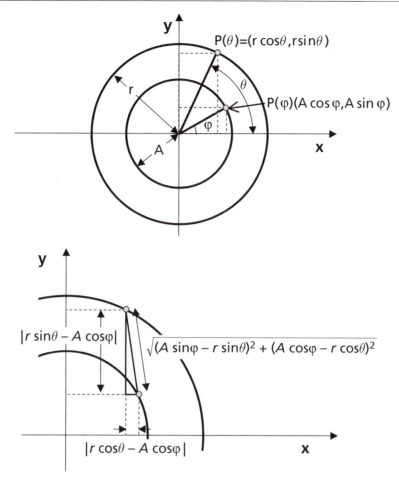

**Figure 5-10. Distance between points on two different circles.**

and the magnitude of the opposite side (*y*-axis) is

$$\left|r\sin\theta - A\sin\varphi\right|.$$

Hence, the distance between $P(\theta)$ and $P(\varphi)$, which we'll write as $\overline{P(\theta)P(\varphi)}$, is found by taking the square root of the **trigonometric distance equation**, which is:

$$\overline{P(\theta)P(\varphi)}^{\,2} = (r\cos\theta - A\cos\varphi)^2 + (r\sin\theta - A\sin\varphi)^2.$$

As a quick set of exercises:

1) What happens if $A = 0$ ?

2) What happens if $\theta = \varphi$, but $r \neq A$ ?

3) What happens if $\theta \neq \varphi$, but $r = A$ ?

Exercise 1) What happens if $A = 0$?

If the circle described by $A$ has zero radius, then the circle is the point of the origin. The trigonometric distance equation reduces to $(r\cos\theta)^2 + (r\sin\theta)^2 = r^2$ which is the equation for the distance from the origin to the point where hypotenuse of the line describing the angle $\theta$ hits the circle of radius $r$.

Exercise 2) What happens if $\theta = \varphi$, but $r \neq A$?

Here we just continue to travel along the hypotenuse from circle of radius $r$ to circle of radius $A$. The distance between the two points is $|r - A|$. We used the absolute value signs as we have not specified which circle is larger.

Exercise 3) What happens if $\theta \neq \varphi$, but $r = A$?

Now we are looking for the chord length between two points on the same circle. With $r = A$ the distance equation reduces to

$$\overline{P(\theta)P(\varphi)}^2 = [r(\cos\theta - \cos\varphi)]^2 + [r(\sin\theta - \sin\varphi)]^2$$
$$= r^2[(\cos\theta - \cos\varphi)^2 + (\sin\theta - \sin\varphi)^2]$$

We'll simplify this equation in the next section.

Please note that the circles and the triangle of which $\overline{P(\theta)P(\alpha)}$ is the hypotenuse are artifices of our procedure to obtain this equation. Actual circles and knowledge of the angles in the newly created triangle are not needed to do our work.

## Adding Trigonometric Functions

Angles add just like other sets of numbers. If we wanted to add angle $\theta$ to angle $\alpha$, the result would be angle $\theta + \alpha$. Likewise, if you wanted to add the value for $\cos\theta$ to that of $\cos\alpha$, the result would be $\cos\theta + \cos\alpha$. These are all numbers and are handled just like any number. To find the cosine of angle $(\theta + \alpha)$ would be an entirely different matter, however. Consider $\cos\frac{\pi}{2}$ which has the value of zero. If we wanted $\cos\left(\frac{\pi}{2} + \frac{\pi}{2}\right)$, we know that the result is $\cos\pi$, which is minus one, not two times zero. What do we do to get trigonometric values when we want to combine angles? The easiest approach is to show how we get the cosine of $(\theta - \alpha)$, and from that proof the others will follow.

Look at Figure 5-11 and use the distance equation to determine the distance from $P(\theta)$ to $P(\alpha)$. The distance equation is simpler now since $r = A = 1$.

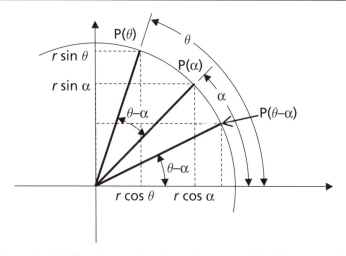

**Figure 5-11. The determination of cosine of additive angles.**

$$\overline{P(\theta)P(\alpha)}^2 = (\cos\theta - \cos\alpha)^2 + (\sin\theta - \sin\alpha)^2$$
$$= \cos^2\theta - 2\cos\theta\cos\alpha + \cos^2\alpha + \sin^2\theta - 2\sin\varphi\sin\alpha + \sin^2\alpha$$

Since we've already shown that

$\cos^2\theta + \cos^2\theta = 1$ (and, of course, the same holds for $\alpha$ )

we simplify this distance equation as

$$\overline{P(\theta)P(\alpha)}^2 = 2 - 2\cos\theta\cos\alpha - 2\sin\theta\sin\alpha .$$

"How does this calculate $\cos(\theta - \alpha)$?" you might ask.

Look now at the arc along the circle from (1,0), which we'll call $P(0)$, to $P(\theta - \alpha)$. That arc is the same length as the arc from $P(\theta)$ to $P(\alpha)$. Hence the chord (or straight line distance) is the same length between each of the two sets of points.

Let's calculate $\overline{P(0)P(\theta - \alpha)}^2$ and equate it to $\overline{P(\theta)P(\alpha)}^2$ and see what we get.

$$\overline{P(0)P(\theta - \alpha)}^2 = (\cos 0 - \cos(\theta - \alpha))^2 + (\sin o - \sin(\theta - \alpha))^2$$

since

$\cos 0 = 1$

and

$\sin 0 = 0$

$$\overline{P(0)P(\theta - \alpha)}^2 = (\cos(\theta - \alpha) - 1)^2 + \sin^2(\theta - \alpha)$$
$$= 2(1 - \cos(\theta - \alpha)) \qquad .$$

The two chords in Figure 5-11 are equal. That is,

$$\overline{P(0)P(\theta-\alpha)}^{\,2} = \overline{P(\theta)P(\alpha)}^{\,2}$$

and then

$$2 - 2\cos(\theta-\alpha) = 2 - 2\cos\theta\cos\alpha - 2\sin\theta\sin\alpha$$

Hence,

$$\cos(\theta-\alpha) = \cos\theta\cos\alpha + \sin\theta\sin\alpha \,.$$

Now that we have this trig identity, there's a lot we can do with it.

## Examples:

1) We've already demonstrated that the cosine of the complement of an angle is equal to the sine of that angle. Let's verify that statement again. Let $\theta = \pi/2$ and use the relationship $\cos(\theta-\alpha) = \cos\theta\cos\alpha + \sin\theta\sin\alpha$ to get

$$\cos(\frac{\pi}{2}-\alpha) = \sin\alpha \,.$$

We've shown this relationship before using a geometrical argument, but it's nice to see that we get the same answer using another approach.

2) Let $\alpha = \dfrac{\pi}{2} - \beta$, put it into the last equation, and we'll show that the sine of the complement of an angle is equal to the cosine of that angle. That is,

$$\sin(\frac{\pi}{2}-\beta) = \cos(\frac{\pi}{2}-[\frac{\pi}{2}-\beta])$$
$$= \cos\beta$$

which we'd also shown earlier in this chapter using a geometric argument.

3) Let's look at the symmetry of sine and cosine about the zero angle. That is, we want to know if $f(-\theta) = f(\theta)$.

3a) The sine:

Let $\alpha = -\beta$ in example (1) above.

Then

$$\sin(-\beta) = \cos[\frac{\pi}{2}-(-\beta)]$$
$$= \cos[\frac{\pi}{2}+\beta]$$
$$= \cos[\beta-(-\frac{\pi}{2})]$$

We then use the recently derived equation for finding the cosine of the difference of two angles to get

$$\sin(-\beta) = \cos[\beta - (-\frac{\pi}{2})]$$

$$= \cos\beta\cos(-\frac{\pi}{2}) + \sin\beta\sin(-\frac{\pi}{2})$$

Since

$$\cos(-\frac{\pi}{2}) = 0$$

and

$$\sin(-\frac{\pi}{2}) = -1$$

then

$$\sin(-\beta) = \cos[\beta - (-\frac{\pi}{2})]$$

$$= -\sin\beta$$

This says that the sine function is an **odd function**. It changes signs when $\beta$ is replaced with $-\beta$. We can also say that the sine is asymmetric with the origin. Of course, you can see this relationship geometrically.

3b) Now let's look at the symmetry of the cosine function. Go back to example (1) and let $\theta = 0$. Then we have

$$\cos(-\alpha) = \cos 0\cos\alpha - \sin 0\sin\alpha$$

$$= \cos\alpha$$

The cosine is an **even function**. It doesn't change signs when $\alpha$ is replaced with $-\alpha$. We can also say that the cosine is symmetric with the origin.

Now that we've covered some basic subtraction identities, let's try addition.

If you want to know how to write $\cos(\theta + \alpha)$ in simpler terms, just rewrite it as $\cos(\theta + \alpha) = \cos[\theta - (-\alpha)]$ and use the subtraction identity to get:

$$\cos(\theta + \alpha) = \cos\theta\cos(-\alpha) + \sin\theta\sin(-\alpha).$$

Then from the symmetry properties of cosine and sine, you have

$$\cos(\theta + \alpha) = \cos\theta\cos\alpha - \sin\theta\sin\alpha,$$

which is another identity that we suggest you commit to memory.

Enough on cosines; let's do the sine.

The addition identity for $\sin(\theta + \alpha)$ is found by using the basic relationship between sines and cosines, which is $\sin\theta = \cos(\frac{\pi}{2} - \theta)$.

So let's write $\sin(\theta + \alpha)$ as

$$\sin(\theta + \alpha) = \cos[\frac{\pi}{2} - (\theta + \alpha)]$$

$$= \cos[(\frac{\pi}{2} - \theta) - \alpha]$$

and we're back to the cosine subtraction identity. We'll continue as

$$\sin(\theta + \alpha) = \cos[(\frac{\pi}{2} - \theta) - \alpha]$$

$$= \cos(\frac{\pi}{2} - \theta)\cos(-\alpha) - \sin(\frac{\pi}{2} - \theta)\sin(-\alpha)$$

$$= \sin\theta\cos\alpha + \cos\theta\sin\alpha$$

We used the symmetry properties in that last step.

Trig identities go on forever. We'll look at a few more and then move on to more interesting stuff.

Look at $\cos(2\theta)$. It's the same as $\cos(\theta + \theta)$, which you know how to simplify to get

$\cos(2\theta) = \cos^2\theta - \sin^2\theta$, and since we know that $\cos^2\theta + \sin^2\theta = 1$, then

$\cos(2\theta) = 2\cos^2\theta - 1$,

which is the same thing as

$\cos(2\theta) = 1 - 2\sin^2\theta$.

Of course, you can do the same type of simplification with $\sin(2\theta)$.

If you feel that we left tangent, secant, cosecant, and cotangent out in the cold, you're both right and wrong. Since

$$\tan\theta = \frac{\sin\theta}{\cos\theta}, \quad \sec\theta = \frac{1}{\cos\theta}, \quad \csc\theta = \frac{1}{\sin\theta}, \text{ and } \cot\theta = \frac{1}{\tan\theta}$$

you can derive any identities that you need from those relationships. Have fun.

We'll relist the three really important identities for your convenience. If you keep these equations and the symmetry properties at ready reference, you should be able to derive any other identities that you should ever want or need.

$$\cos^2\theta + \sin^2\theta = 1$$
$$\cos(\theta \pm \alpha) = \cos\theta\cos\alpha \pm \sin\theta\sin\alpha$$
$$\sin(\theta \pm \alpha) = \sin\theta\cos\alpha \pm \cos\theta\sin\alpha.$$

## Summing Interior Angles of Triangles and Quadrangles

Before doing the triangle, let's look at the simplest quadrangle[12], the right rectangle as in Figure 5-9. Since the vertical lines are perpendicular to the horizontal lines, we can assume that all interior angles of that rectangle are 90°. Consequently, the sum of the interior angles in a right rectangle equals $2\pi$ radians, or 360°.

Now, draw a dotted line from one corner of the rectangle diagonally to the opposite corner. You have two identical, but rotated, triangles. The values of the angles $\theta$ and $\alpha$, as shown in Figure 5-12, depend on the relative lengths of the horizontal and vertical sides, but their individual values are not important to this discussion. What is important is that the $\theta$s and $\alpha$s in both corners are the same and that $\theta$ plus $\alpha$ equals 90°. With the remaining 90° angles at the unaltered corners, the interior angles of the right triangle equals 180° or $\pi$ radians.

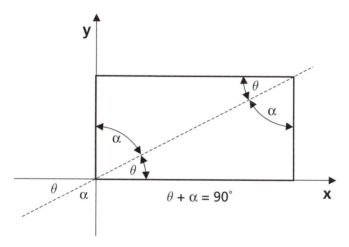

**Figure 5-12. Interior angles of a right triangle.**

What about a general triangle?

That's easy, but before going general, let's look at an isosceles[13] triangle. We can make an isosceles triangle by extending a mirror image across one of the legs of a right triangle as shown in Figure 5-13. All of the $\theta$s and $\alpha$s are the same values as before, and we have the sum of interior angles = $2(\theta + \alpha)$.

---

[12] A general quadrangle is a four-sided figure without specified angles.

[13] A triangle with two equal legs is an isosceles triangle. Special names for other triangles are: acute, all three angles less than 90°; obtuse, one angle is greater than 90°; equilateral, all legs are same length; and scalene, where no two legs are equal. Note that the isosceles and scalene might also be right triangles. The obtuse and equilateral cannot be right triangles.

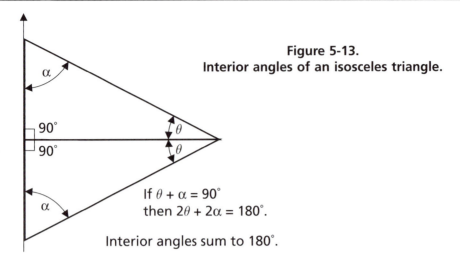

**Figure 5-13.
Interior angles of an isosceles triangle.**

90°

90°

$\theta$

$\theta$

$\alpha$

$\alpha$

If $\theta + \alpha = 90°$
then $2\theta + 2\alpha = 180°$.

Interior angles sum to 180°.

We've already convinced you that $\theta + \alpha$ equals 90°, so:

the sum of interior angles = 180°.

If you want to be even more general and have all three sides of the triangle be different lengths, just draw a rectangle and add dotted lines from adjacent corners on the lower boundary to an arbitrary position on the upper boundary as in Figure 5-14. You now have a general triangle which you can turn into two right triangles by dropping a vertical line from the arbitrary meeting place. You now have four different interior angles to sum to get the interior angles of your general triangle. If you identify the exterior angles, you'll see the 180° rule still holds.

Now let's return to four-sided figures. Go back to the rectangle. Fix the lower horizontal boundary and push the upper edge to the right and you make a parallelogram as in Figure 5-15. Again, drop lines to make right triangles, identify similar angles, and sum up the interior angles. They still sum to 360°.

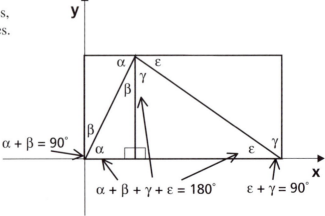

**Figure 5-14. Interior angles of a general triangle.**

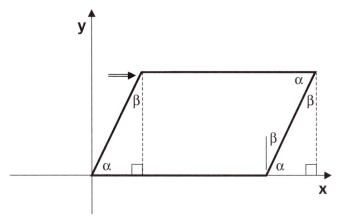

**Figure 5-15. Interior angles of a parallelogram.**

If you want to show that the interior angles of the general quadrangle also sum to 360°, you'll have to do the work on this one, but we'll get you started. Choose one of the sides as your base. Make a rectangle about the whole quadrangle and then drop lines making right triangles. You have the tools to continue, so do it.

## The Laws of Sines and of Cosines

We will do two more laws or relationships, and then we'll move on to more interesting material. Consider a general triangle which we've turned into two right triangles as in Figure 5-14, now relabeled in Figure 5-16. The vertex[14] of the triangle, which we'll call $H$, is equal to both $A\sin\theta$ and $T\sin\alpha$. Hence

$$\frac{\sin\alpha}{A} = \frac{\sin\theta}{T}.$$

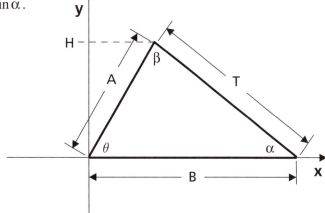

**Figure 5-16. Development of the Law of Sines and Cosines.**

---

[14] Vertex has a double meaning. If we are only talking about an angle, then the vertex is the point where the two lines meet. If we are talking about a triangle or a mountain, then the vertex is the angle (or peak) opposite to and farthest from the base.

We can do the same manipulation with any two sets of angles within the triangle. Thus,

$$\frac{\sin\alpha}{A} = \frac{\sin\theta}{T} = \frac{\sin\beta}{B}.$$

And this is the **Law of Sines**.

Let's not leave out cosines. The vertex of angle $\beta$ is at coordinates $(A\sin\theta, A\cos\theta)$. Hence, the length of $T$ is:

$$T^2 = (B - A\cos\theta)^2 + (A\sin\theta)^2$$
$$= B^2 - 2AB\cos\theta + A^2\cos^2\theta + A^2\sin^2\theta.$$
$$= A^2 + B^2 - 2AB\cos\theta$$

This is the **Law of Cosines**, or a re-statement of the Pythagorean theorem for a general triangle.

## Example:

Apollo astronauts left laser-beam reflectors on the surface of the moon. An Earth-based telescope focuses the already highly collimated laser beam and aims at the reflector. Then it receives the return signal approximately 2.4 seconds later. Even though the laser has the lowest angular dispersion of any known light source, the beam spreads to a four-mile width at the surface of the moon.

*Question*: Ignoring atmospheric effects, what is the angle of the laser beam?

*Answer*: We have a triangle where one leg is four miles long and another is the distance from the Earth to the moon. The third leg is so close to the same length as the Earth-moon leg that we ignore the difference. And implicit in choosing this triangle is that the laser beam is a point at its origin.

We need the distance from the Earth to the moon to complete this problem. You know that the speed of light in a vacuum is $3 \times 10^8$ meters/second, so the Earth-moon distance is

$$D = vt$$
$$= (3 \cdot 10^8 \frac{m}{s})(\frac{2.4s}{2})$$
$$= 3.6 \cdot 10^8 \, m$$

or approximately 360,000 kilometers.

But, we have a units problem. The leg of the triangle on the moon is given in miles. The only metric-to-English conversion we remember is that one inch is identically equal to 2.54 centimeters. We must convert miles to centimeters and then convert centimeters to kilometers. Easy, just messy.

4 miles = 4 miles × 5280 feet/mile × 12 inches/foot × 2.54 centimeters/inch

= 643,737.6 centimeters or approximately 6.4 kilometers[15].

Since the angle is so small, we can look for either the sine or the tangent of the angle and be well within the accuracy of our measurements. So the beam divergence, $\theta$, is:

$$\arcsin\left(\frac{6.4}{360,000}\right) \approx 1.8 \cdot 10^{-5} \text{radian}$$

$$\approx 0.001 \text{ degree}$$

and that's a very narrow beam.

# Periodicity

Periodicity may sound boring. It may sound like the same old thing over and over—like, where's the excitement? However, the fun is there because we are talking about turning trig functions into waves. We'll modify these waves and combine them to do interesting things. We can even use these waves to explain why that great surfing wave only comes along every once in a while.

Before we get into more trig functions, let's define a periodic function. Simply stated, a *periodic function* is one that has the form:

$$f(x) = f(x + n\lambda)$$

where

$n$ = any positive integer

and

$\lambda$ = a constant with the same units as $x$[16].

This function $f(x)$ repeats itself in every period of $\lambda$.

Now, let's get back to trig. We'll start with the sine function and hope that we've convinced you that

$\sin\theta = \sin(2n\pi + \theta)$, where $n$ is any positive integer.

---

[15] We have no business converting a value given to one decimal place (4 miles) to seven places (643,737.6 centimeters) just because the conversion factors pushed us that way. When a value is stated as simply 4, we may well assume that the true value is between 3.5 and 4.4. If we want precision to six places then our starting value should be 4.000000, which means that we know the value to better than one part in a million.

[16] Please be very careful with $x$ and $y$ as we use them here. Our prior use of $x$ and $y$ in this chapter was to define a coordinate system in which to draw a circle. The variable that we played with was the angle $\theta$ or some other angle as the argument of the trig functions. We are now going to put $x$ inside of the trig function. This is *not* the same $x$. Why not use another symbol for our variable you might ask? Tradition is our only answer.

The period of this sine function is $2\pi$ radians.

We've talked about the repeating values of the trig functions as we go around the circle that covers $2\pi$ radians. If we want to look at these trig functions as they repeat their values, and especially if we want to combine trig functions or otherwise modify them, life is easier if we unfold the circle. Instead of going around the circle and trying to keep track of which loop we might be on, we'll plot the amplitude of the function on the $y$-axis and the angle, $\theta$, on the $x$-axis, as shown in Figure 5-17.

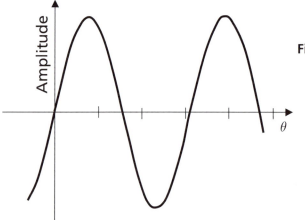

**Figure 5-17. A sine wave.**

We can plot $\sin\theta$ vs. $\theta$, but it's messy since things happen at integers and fractions of $\pi$, and that makes labeling the $\theta$-axis more difficult than it needs to be. Let's redefine $\theta$ as

$$\theta = \frac{2\pi}{\lambda}x$$

so

$$\sin\theta = \sin(\frac{2\pi}{\lambda}x)$$

which we can plot as in Figure 5-18. Maximas, minimas, and axis crossings now occur at simple fractions or integers of $x/\lambda$. Life is easier. We see that $\lambda$ is the length[17] of the period. It is often called the wavelength (or periodicity) of whatever you are observing. And please note that our new variable, $\frac{2\pi}{\lambda}x$, must be dimensionless, just as $\theta$ had no dimensions. That is, $x$ and $\lambda$ must have the same units.

---

[17] We use the term length somewhat casually. Length here is not necessarily linear distance. It is the magnitude of our variable that gives the function its periodicity. If we are talking about ocean waves, then linear distance is a good choice, but other phenomena are better described using other units. We'll soon describe ocean waves with a periodicity in time and then in terms of distance and time.

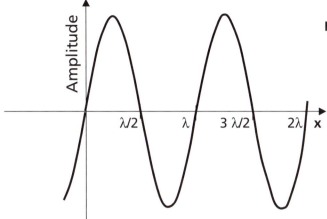

**Figure 5-18. Sine wave plotted in terms of its wavelength.**

Let's add another interesting and useful feature to our sine function. We have the function to describe a stationary wave—that is, one that does not change with time. For your next assignment, go to the beach, wade into the water out past the breakers, and watch the waves roll by. Compare the shapes of the undulating waves with Figure 5-18 and see if the plotted sine function (a sine wave) isn't a reasonably good approximation for describing the observed physical phenomena. The distance between waves is $\lambda$. If your beach is like ours, you'll always think of trigonometry when you go surfing.

Sine waves roughly[18] describe ocean waves except for one very important feature. The sine wave as we've shown it is stationary. Ocean waves move. If you look out to sea and make a snapshot of the waves, they are stationary for that instant, and your function works. For your next assignment, while still in the water don't look out to sea, but observe the waves as they roll pass by you. The water level is going up and down as a function of time. Hold a ruler and make notes of the changing water level while observing the time on your water-proof watch, and we bet you can fit this data to an equation such as

$$h(t) = H \sin(2\pi f t)$$

where $H$ is the average wave height and $f$ is a constant, which we'll call frequency[19]. It is the number of times per unit time that you observe a peak wave height. Note that $f$ has the units of reciprocal time.

---

[18] We'll show you some stuff later in the chapter which allows you to alter your sine waves to better describe the ocean waves.

[19] The frequency is often stated in radians per unit time and uses $\omega$ as its symbol. $\omega = 2\pi f$. We'll stay with $f$ for now.

If we plot this function, it'll look very similar to Figure 5-18 except the independent variable is time rather than distance, and we have *f* instead of $\lambda$ in the argument.

As we said, $\lambda$ is the wave length if we take a snapshot. The corollary to $\lambda$ in the time-dependent function is *f*, the frequency, which is the number of times the sine function repeats itself in a unit of time[20].

We don't have to choose between taking snapshots and only looking at changes at a specified point. We can combine the two into a function of space and time as

$$h(x,t) = H\sin(2\pi\frac{x}{\lambda} - 2\pi ft) \, .$$

The sign indicates whether the wave is going towards the left or towards the right. Switch the sign and the wave will go in the other direction. How can you tell whether it should be a plus or minus sign? The easiest way is to plot a couple of points. If you prefer a more mathematical approach, then ask the question: for a constant $h(x,t)$, does *x* need to increase or decrease as time increases? By looking at the equation, we see that *x* would need to increase with respect to time to keep

$\sin(2\pi\frac{x}{\lambda} - 2\pi ft)$ constant.

And now we can pull more information out of our trig function. Our function gives us the wavelength of our observed phenomena and the number of times that phenomena occurs at some chosen point per unit time. That is, we have length and time of the separations of the peaks of the waves, so we have velocity.

For example, get a colleague to join you in the surf. Have your fellow scientist separate from you in the surf until each of you is standing (Figure 5-19) far enough apart to measure the distance between waves. You have measured the wavelength. With your waterproof watch, measure the time period between waves. The reciprocal of that time period is the frequency of the waves. Velocity, *v*, is distance divided by time so

$$v = \lambda f$$

or

$$v = \frac{\lambda\omega}{2\pi} \, .$$

---

[20] If the time unit is seconds then the proper term for *cycles per second* is hertz. That is, you'd state the frequency of 1000 cycles per second (or 1000 cps) as 1000 hertz (Hz). This nomenclature honors the nineteenth-century German physicist Heinrich Rudolf Hertz, the first person to generate what we now call radio waves.

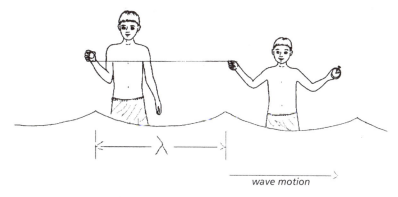

**Figure 5-19. Sketch of two mathematicians researching
the velocity of ocean waves.**

Typical measurements are $\lambda = 12$ meters and $t = 6$ sec between waves. So what is the frequency of the waves and how fast are they traveling?

## Phase Differences

Normally we can start measuring our sine waves when and where we want to. If, however, we are trying to compare two different waves of the same frequency starting at different times we need another tool in our box. If they happened to start at the same time and travel the same path, we would just add their amplitudes. If our second sine wave started at some other time, as in Figure 5-20, then there would be a **phase difference** between the two waves. We could say that the second wave has a **phase shift** relative to the first one. We write this new function with the phase shift as

$$f(t) = A \sin( 2\pi f t + \varphi )$$

where $\varphi$ is our phase shift.

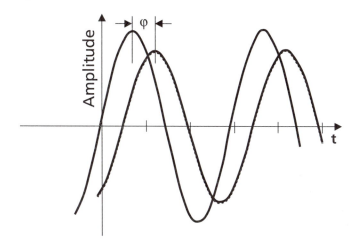

**Figure 5-20. Phase differences in waves.**

The phase shift tells us mathematically that the sine wave started earlier if $\varphi > 0$, or later if $\varphi < 0$. A phase shift between waves can also occur when a single wave splits due to some interference, and the two different waves follow different paths only to recombine at a later time.

We can also talk about phase differences in the spatial domain of our wave functions. We can easily have two waves that are out-of-phase arriving at some point. In that case we'd express one wave as

$$f(x) = \sin(2\pi \frac{x}{\lambda})$$

and the other as

$$f(x) = \sin(2\pi \frac{x}{\lambda} + \varphi)$$

A simple example of this phenomena[21] is found in the study of x-ray diffraction in determining crystal structure. Look at Figure 5-21. A narrow beam of monochromic x-rays[22] impinges upon the surface of a crystal. Part of the beam is reflected off of the atoms at the surface of the crystal, and part of the beam penetrates the surface to be reflected from lower layers of atoms within the crystal. The waves reflected at the surface combine with waves that emanated earlier from the x-ray source which traveled into, then out of the crystal. The combined waves[23] then traveled on to the detector. The waves started from the same source and were in phase when they began their trip. They are now out-of-phase, but by how much?

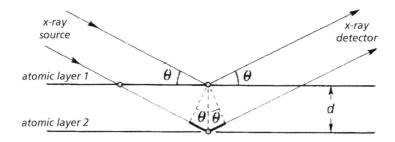

**Figure 5-21. Sketch of x-ray diffraction and Bragg's Law.**

---

[21] It may sound simple now but the father and son team of William L. and William H. Bragg won the 1915 Nobel Prize for this discovery.

[22] A typical x-ray generator produces a broad spectrum of x-rays. Proper selection of x-ray targets and filtering allow us to use x-rays of only a single wavelength, hence the term monochromic.

[23] Since these two waves are now combined, some folks would consider them one wave and use the *singular*.

The two waves traveled the same distance except for the scenic route into and back out of the crystal. That extra distance is $2d$ (twice the separation of the crystal planes) if the sources and detector were at right angles to the crystal surface. Otherwise, the extra distance is $2d \sin\theta$, where $\theta$ is the angle between the x-ray beam and the surface of the crystal.

So what does this have to do with phase? Lots. The penetrating wave has traveled an extra distance of

$$2d \sin\theta$$

which is a portion of the length of the waves of the radiation. Hence, we have a phase change between the two waves equal to

$$\varphi = \frac{2d \sin\theta}{\lambda} 2\pi$$

The waves combining and producing a signal in the detector are

$$f(x) = \sin(2\pi\frac{x}{\lambda})$$

and $f_2(x) = \sin(2\pi\frac{x}{\lambda} + \varphi)$

where $f_1(x)$ is the beam reflected at the surface

and

$f_2(x)$ is the beam reflected from within the crystal.

If the phase change equals an integral number times $2\pi$, the waves reflected from those two layers and subsequent layers deeper within the crystal will add and give you a nice fat signal at the detector. Otherwise, the waves will destructively interfere with each other and not be detected.

If you know the wavelength of your x-rays, you can determine the distance[24] between layers of atoms in your crystal from the Bragg equation:

$$d = \frac{n\lambda}{2\sin\theta}, \text{ where } n \text{ is an integer,}$$

---

[24] You can determine the atomic structure of your crystal as we have indicated. You can identify materials by measuring the atomic spacing as we have shown and comparing your measurements with data from previous measurements. You can also identify the elements within a material by making that material produce x-rays (by hitting it with a strong x-ray beam or a beam of high-energy electrons) and measuring the wave lengths of the newly created x-rays using a crystal of known atomic spacing and solving for $\lambda$ in the Bragg equation.

which honors the father-son team who originated it[25].

Figure 5-22 shows the detector output for CuK $\alpha$ radiation and tungsten crystals.

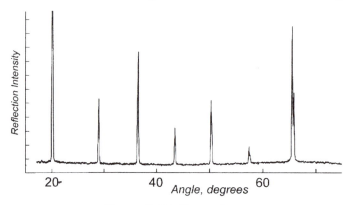

**Figure 5-22. X-ray data.**

# Effect of Combining Waves with Small Frequency Differences

We have just shown the effect of combining waves of the same frequency but with a phase difference. The result is static—it repeats itself without variation. Combining waves of slightly different frequencies[26] can produce some startling effects.

Did you ever wonder why the good surfing waves tend to come in groups with lots of small waves in between? It's because of slight frequency shifts in the sine waves. Dream that you are sitting on your surf board at Waikiki Beach. The waves are rolling in. Some are good riders, and some just give you a gentle sway, putting you to sleep. To stay awake you start taking measurements, and you tune your wrist computer to pick up the NOAA web site. A storm in the Sea of Japan is kicking up one set of waves, a mild earthquake near Alaska is producing another, and a convoy of tankers just off shore is making a third set of waves. Something about ocean dynamics makes the frequency of the waves close to the same value. Close, but not exactly the same, and you are in for some fun.

---

[25] A similar argument produces equations that explain the colors in such diverse things as an oil slick, a soap bubble, and the iridescence seen in some bird's feathers. However, the explanation for soap bubbles is more complicated than that of x-ray diffraction. The light wave travels an extra distance into and out of the bubble layer just as with x-rays and crystals, but with bubbles, the wave inverts itself upon the first reflection, it changes speed within the soap, and you must consider many wavelengths within the visible light.

[26] Wait until we get into the calculus chapter. We'll show you how to create almost any waveform by combining sine and cosine waves. We call those operations Fourier analysis.

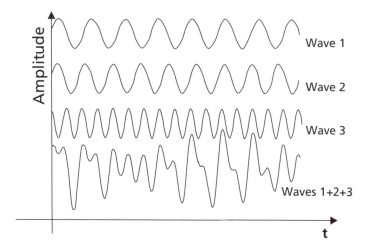

**Figure 5-23. Combining waves of different frequencies.**

Figure 5-23 shows the three waves from those different sources. We add those waves, and you see where the good riders come from. Imagine all of the possible sources of wave disturbance and formation in that beautiful ocean, and you'll understand how those really big ones sneak up on you.

## Dissipation

So far we've talked about waves as though, once formed, they go on forever. They don't. And that gives us an excuse to entertain you with another functional form: the dissipative wave. It matters not if you are talking about sound waves, waves formed by throwing a rock into a puddle, or electromagnetic waves created by discharging a capacitor through a coil[27]. All of these waves diminish in amplitude with time or with distance from the source. The dissipation factor is normally a negative exponential function[28].

For an example, we'll look at a capacitor-coil system that has a small amount of electrical resistance. We only need the time-dependent equation since we are reading a voltage between two points in our circuit with an oscilloscope.

The describing equation will have the form

$$v(t) = Ve^{-\alpha t} \sin(2\pi f t + \vartheta)$$

---

[27] No superconductivity allowed in this example.

[28] No, we're not making this up. You'll see why in the differential equations chapter.

where

V is the voltage to which the capacitor was initially charged

$\alpha$ is dissipation factor and is equal to $\dfrac{R}{2L}$

$f$ is the frequency of oscillation[29] and is equal to $\dfrac{1}{2\pi\sqrt{LC}}$

and

$\vartheta$ is a phase angle that depends on where we place the oscilloscope's leads,

R is the resistance in ohms,

L is the coil's inductance in henrys

and

C is the capacitance in farads.

Figure 5-24 shows this, where we chose circuit values $C = 2 \times 10^{-6}$ farads, $L = 1.5 \times 10^{-3}$ henrys, and R = 40 ohms.

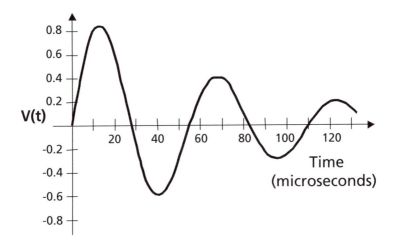

**Figure 5-24. Wave dispersion of a coil-capacitor system.**

---

[29] The frequency is modified by the dissipation factor, but for the values that we chose, the change is small. More details will follow in differential equations.

# Vectors

We have talked about angles and trig functions. The angles as we have drawn them have legs of different lengths and different angle relationships among the lines. The trig functions have an amplitude and also have frequency (or wavelength) and phase relationships among various functions.

So what are leading to? Vectors. A **vector** is a representation of both a quantity and a direction. That's a mouthful. If we're only talking about things happening in a spatial coordinate system, the vector represents both a magnitude of a phenomena, such as wind speed, force, and such, and the direction in which that phenomena is acting.

Let's look at Newton's second law to demonstrate the beauty of vectors. This law, to which there is no appeal[30], says that an object will gain speed[31] proportional to the sum of forces on that object. An apple suddenly becomes detached from the tree. It falls faster and faster as time passes during its flight, and, *voila*, the force on gravity is discovered. Wasn't the force of gravity pulling on the apple while it was still attached? Of course it was, but there was an equal force on the apple in the opposite direction exerted by the branch. If we call the magnitude of gravity's force on the apple $F_{gravity}$ and the magnitude of the force exerted by the branch on the apple $F_{branch}$ then, before breaking:

$$F_{gravity} = F_{branch}.$$

This may seem strange. With twice the magnitude of force on the apple, it doesn't fall. Cut the magnitude in half, and the apple falls. The key to this enigma is the word "magnitude." The force of gravity is pulling the apple down. The force of the branch is pulling up. Force is a vector quantity; we must consider direction. Equal and opposite forces cancel out, and the speed of the object does not change.

Consider three people pushing a truck stuck in the mud. If they cooperate and all push in the same direction, all is well. If, however, they push in different directions, they'll stay in the mud.

If you like these vectors, we have a real treat for you in the chapter on vector calculus.

## Examples

### Example 1: Statics

Press your palms together in front of your body, so your left hand is exerting a force to the right, and your right hand is exerting a force towards the left. You can feel the force in your arms. If you push equally hard with each hand, nothing moves.

---

[30] Though modified by quantum mechanics.
[31] Such a gain in speed is called acceleration. We'll see more of it in Chapter 6.

This observation is vital for civil engineers to understand if they want their buildings to remain in one place: *the sum of forces at any place and at any time must equal zero.* If the forces do not sum to zero, the building will move because force equals mass times acceleration. The building is the mass, and the acceleration of that mass is something we don't want.

Look at Figure 5-25. The cable is 10 meters long and the ceiling hooks are separated by 8 meters. The box is in the center of the cable. Using the civil engineer's law for stationary buildings, determine the force in the cable in Figure 5-25 if the box weighs 100 newtons.

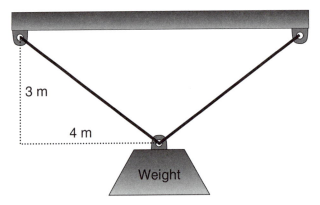

**Figure 5-25. Diagram of a weight supported in the center of a cable connected at both ends to an overhanging beam.**

First we'll draw what engineers call a free-body diagram, as shown in Figure 5-26.

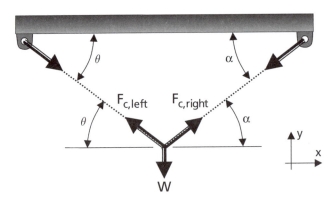

**Figure 5-26. Free-body diagram of a weight supported at the center of a suspended cable.**

As a cable cannot support a shear force, all force must be transferred along the cable to the ceiling supports, but the force in each leg of the cable is greater than half the total weight of the box. Imagine for a moment that the ceiling hooks were free to move along the ceiling: if so, they would slide towards each other. They don't move. Hence, there is a lateral force on these hooks.

We can determine the force in the cable two different ways. With both techniques we need to identify the angle that the cable makes with the ceiling. We'll call the angle on the left $\theta$ and the angle on the right we'll call $\alpha$. In this particular case $\theta = \alpha$ since the weight is suspended at the center of the cable.

**Method #1.** Sum the forces in each direction.

$$\sum F_x = -F_{c,left} \cos\theta + F_{c,right} \cos\alpha = 0$$

and

$$\sum F_y = F_{c,left} \sin\theta + F_{c,right} \sin\alpha - W = 0 .$$

where $F_{c,left}$ is the force in the left-hand side of the cable.

You have two equations and two unknowns. Even if $\theta \neq \alpha$ you could still solve for the force in each leg of the cable.

In this case, the summation of the $x$-forces tells us that $F_{c,left} = F_{c,right}$ (which we'll call $F_c$), and summation of the $y$-forces tells that

$$F_c = \frac{100}{2\sin\theta}$$
$$= 50 * \frac{5}{3}$$
$$= 83.3$$

If you're interested, you could also calculate the lateral ($x$–direction) force on the ceiling hooks. And that is

$$F_{x,left} = F_{c,left} \cos\theta$$
$$= 83.3 * \frac{4}{5}$$
$$= 66.7$$

Where, of course, the force is expressed in newtons.

**Method #2.** Law of sines.

Remember the law of sines? If we connect the two vectors representing the force in the two legs of the cable with the vector representing the weight (and they must connect since we have a static system), we'll have the relationship

$$\frac{100}{\sin(\alpha+\theta)} = \frac{F_{c,right}}{\sin(90-\theta)}$$

$$= \frac{F_{c,left}}{\sin(90-\alpha)} .$$

Look at Figure 5-27 to sort out the angles.

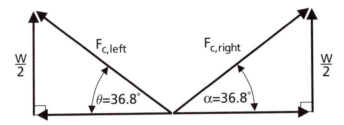

**Figure 5-27. Solving the force in the cable using the law of sines.**

We know from the symmetry of the problem that $\theta = \alpha$ so

$$F_c = 100 * \frac{\sin(90-\theta)}{\sin(\alpha+\theta)}$$

$$= 100 * \frac{\cos\theta}{\sin(2\theta)}$$

$$= 100 * \frac{0.8}{\sin 73.7} ,$$

$$= 83.3$$

which is the same answer we got with the summation of forces method.

Now let's put our mathematical education to a practical use and play tug-of-war. Go find the strongest person you know, whom we'll call Ox. Assume that Ox is five times stronger than you are, and you are going to challenge him to a game of tug-of-war. It won't be a fair game because he will only use muscle, and you will use muscle and brains.

Tie one end of the rope to a tree, hand Ox the other end, and you grab the middle. The game is to see who can pull the other person two meters. Since you are at the middle of the stretched rope, Ox can't pull you. You can't lose. However, you want to win. You must calculate the required distance between the tree and Ox for you to be able to pull him the required two meters to win the game.

Figure 5-28 shows the set-up. The rope length between the tree and Ox is $x$ meters. The hypotenuse of the triangle at the conclusion of the game is $x/2$ meters.

The base of the triangle is $(\frac{x}{2} - 1)$ meters. Likewise the force in the rope is five times the force in the y-direction. So,

$$\sin\theta = \frac{1}{5}$$

and

$$\cos\theta = \frac{x/2 - 1}{x/2}$$
$$= 1 - 2/x .$$

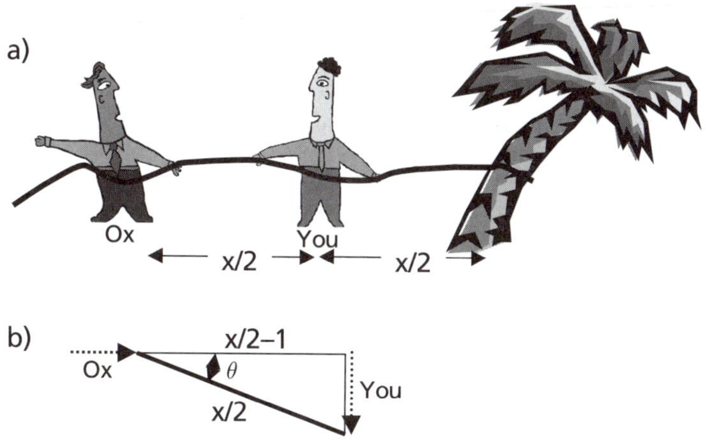

a)

b)

Figure 5-28. Tug-of-War set up.

There are many ways to solve this one, but we'll use the trigonometric identity:

$$\cos^2\theta + \sin^2\theta = 1 .$$

Using our values in the tug-of-war game, we have

$$(1 - \frac{2}{x})^2 + (\frac{1}{5})^5 = 1 .$$

We want to solve for x, the length of the rope between the tree and Ox.

Since the equation above is second order, we have two mathematical solutions which are

$x = 99$ meters

and

$x = 1$ meter.

Since you can't take two meters away from a one meter rope, the second mathematical solution is not a physically realizable solution. Go get your 99-meter rope, plus enough to tie around the tree, and challenge Ox to a game.

## Example 2: Friction

Friction is the resistance to movement of a body that sits upon a surface. The coefficient of friction is the ratio of the force needed to move that body to the weight of the body as

$$\mu_s = \frac{F}{N}$$

where

$\mu_s$ is the coefficient of static[32] friction

$F$ is the force needed to start the body moving

$N$ is the weight of the body that is normal to the surface.

For a very simple example, assume that we have a 30-kilogram block on a horizontal surface, and assume that $\mu_s = 0.3$ (typical for wood on wood). What force must you garner to start the weight moving?

You'll use the equation

$$F = \mu_s N$$
$$= .03 * 30 * 9.8 \ .$$
$$= 88.2$$

Hence we need a force of 88.2 newtons to start the wooden block moving. Since the coefficient of kinetic friction for wood on wood is only about 0.2, our block will accelerate if we maintain the 88.2-newton force. The 9.8 factor is our gravitational factor.

There was a bit of physics in that example but no trig. Let's let trig provide the force to start the block moving. In Figure 5-29 we are going to tilt the wooden surface until the block starts to move. Using the coefficients of static and kinetic friction given above, what angle do we need to start motion and what angle do we want, after motion begins, to maintain movement but prevent acceleration?

---

[32] The coefficient of static friction refers to the force to start motion. Once motion begins, it takes less force to maintain motion, and we refer to this as the coefficient of kinetic friction. Needless to say, different materials have different coefficients of friction.

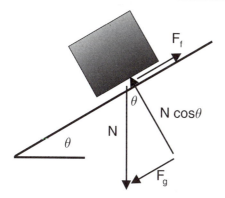

**Figure 5-29. Motion of a block on an inclined plane.**

As we tilt the block, the weight normal to the horizontal surface is no longer $N$, but is $N \cos\theta$, where $\theta$ is the angle between the horizontal and the inclined plane. Not only is gravity reducing the amount of weight on the plane, but it is also providing a force to try to start the motion. Gravity is pulling on our block with the force of $N \sin\theta$. Hence, to start the block moving, we need to know when the retarding force of friction, $F_f$, equals the gravitational force, $F_g$.

The frictional force is

$$F_f = \mu_s N$$
$$= \mu_s W \cos\theta$$

and

$$F_g = W \sin\theta$$

where W is the weight of the block—not just the normal force.

Equating these two equations we get,

$$\mu_s = \frac{\sin\theta}{\cos\theta}$$
$$= \tan\theta \; .$$
$$= 0.3$$

The arctan of 0.3 is approximately 17°. Try it with a couple of smooth boards and see if your experiment confirms our numbers.

Now we want to lower the angle of the surface so that the block continues to move but does not accelerate. We use the same equations but use $\mu_k$ in lieu of $\mu_s$. Hence,

$$\mu_k = \frac{\sin\theta}{\cos\theta}$$
$$= \tan\theta \; .$$
$$= 0.2$$

The arctan of 0.2 is approximately $11°$.

Let's put our knowledge of trig and friction to work for us. Our friend Dufus is helping his friends build a house. He is cutting a sheet of plywood. The upper end of plywood is nailed to the rafters, Dufus is sitting on the lower end of the plywood, and he is cutting across the center of the sheet. The rafters have a 4-12 pitch. Is Dufus in for a ride and a fall when he finishes his cut?

When Dufus finishes sawing the sheet of plywood, it is free to move since only the upper end was nailed to the rafters. As you saw earlier, Dufus's weight is not part of the equation that determines if he slides. Only the angle counts. You must calculate the angle to know if you should call the rescue squad to help put Dufus back together again after he slides off the roof. If you're not a civil engineer or a carpenter, you may not realize that the roof's angle has already been given. The 4-12 pitch is the ratio of *rise-to-run* of the rafters. That is, for every 12 units of width of the house, the roof rises 4 units. Hence the angle is:

$$\arctan \frac{rise}{run} =$$
$$= \arctan \frac{4}{12}.$$
$$= 18.4°$$

As we saw earlier, the wood should start sliding if the angle is $17°$, so Dufus had better move fast[33].

## Example 3: Optics: Reflection and Refraction

The law of reflection is straightforward. As shown in Figure 5-30, the angle of reflection equals (with a convenient change in reference point) the incident angle (that is $\theta_1 = \theta_3$), and both incident and reflected beams are in the same plane. This law can readily be tested with a flashlight and a mirror.

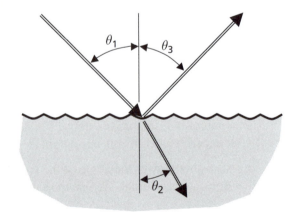

**Figure 5-30. The laws of reflection and refraction.**

---

[33] This is a true story. Only the name was changed to protect one of your authors. The board started moving as soon as the cut was complete. Dufus had a spinning saw in his hand as he rode the plywood toward the edge of a two-story house. He jumped off the board, onto the open rafters and neither dropped the saw nor removed any fingers. Only his pride was injured.

The law of refraction is qualitatively observed when you insert one end of a straight stick into a pool of water: the stick seems to bend[34]. Of course the stick doesn't bend unless the "pool of water" is really a powerful stream that may even pull you in and douse you about. The bending is only illusionary for it's the light beam that bends as shown in Figure 5-30.

Mr. Snell quantified this bending. The law, named for him, is

$$n_1 \sin \theta_1 = n_2 \sin \theta_2,$$

where

$n_i$ is the ratio of the speed of light in a vacuum[35] to the speed of light in material $i$, and it is called the "index of refraction,"

and

the angles are defined in Figure 5-30.

Obviously the index of refraction of a vacuum is 1, and the index of refraction of air is so close to 1 that when we are going between air and most materials, we can take $n_{air}$ as equal[36] to 1. So if we are going between air and some denser material, we can write Snell's law as

$$n = \frac{\sin \theta}{\sin \theta_m}$$

where

$\theta$ is the angle between the beam and a normal vector to the surface on the air side of that surface

and

$\theta_m$ is the angle between the beam and a normal vector to the surface within the material of interest.

A few typical indices of refraction are

### Table 5-2. Index of refraction for some common materials.

| Material | Index of refraction |
|---|---|
| vacuum | 1.0000 |
| air | 1.0003 |
| water | 1.33 |
| typical glass | 1.52 |
| diamond | 2.42 |

---

[34] We call this bending, "refraction." Refraction means "broken" in Latin because the stick appears broken when put into the pool.

[35] Light travels faster in a vacuum than in any material.

[36] But not always. The desert (or highway) mirage is caused by the difference in the index of refraction of the hotter air near the surface and the cooler air away from the surface.

Your dubious friend wants to sell you a large diamond at a low price. Before pulling out your check book you remember Snell's law. You shine a beam of light onto the *diamond* surface at an incident angle of 45° and note that the angle of refraction was approximately 28°. Do you buy the diamond?

You apply Snell's law

$$n = \frac{\sin\theta}{\sin\theta_m}$$
$$= \frac{\sin 45}{\sin 28}.$$
$$= 1.5$$

From the table, you conclude that your ex-friend has a hunk of glass.

Due to your intellectual interest, you find a real diamond and conduct the same test. What angle of refraction do you expect if you use the same angle of incidence?

You go to the table and pick $n_{diamond}$ of 2.42, and put it with $\sin 45$ to get

$$\sin\theta_{diamond} = \frac{\sin 45}{n_{diamond}}$$
$$= \frac{\sin 45}{2.42}.$$
$$= 0.29$$

The angle that you expect to see is the arcsine of 0.29, which is about 17°. Do you suppose that this greater index of refraction, and concomitantly, the larger bending of light rays, has anything to do with the greater luster of diamond over glass?

So far we have been on the outside looking onto the higher index of refraction material. It's a hot day. Let's jump in the deep end of the pool and look[37] out. What do you see? You might see the stick, bent, emerging from the water, and you might not see it.

Don't forget, you are in the water, so the incident angle is now the angle between the normal to the surface within the pool and your viewing direction. If you stick with a 45° incident angle, the angle between your observation and the air side normal is the arcsine of

$$\sin\theta_{air} = n_{water}\sin\theta_{water}$$
$$= 1.33 * \sin 45$$
$$= 0.94$$

which means that $\theta_{air}$ = 70 degrees.

Now, you are running out of air, and you go a bit closer to the surface, but you are still looking at the same stick. That is, you are increasing the angle of incidence. When you increased your viewing angle to 49°, you no longer saw the outside world. For a moment you almost panic and come to the surface, but then you remember your trig and Snell's law.

You put the 49° into Snell's law and determine that you've asked the impossible because

$$\sin\theta_{air} = n_{water}\sin\theta_{water}$$
$$= 1.33 * \sin 49$$
$$= 1.004$$

which, of course, is not a realistic solution. What you discovered is the angle of total internal reflection. If you were shining a light from within the water at that angle, none of the light would leave the surface. All of the light, except that which was absorbed in the water, would be reflected back into the water.

Looking at the result above, you see that the angle of total internal reflection, $\theta_r$, is $\theta_T = \arcsin(\frac{1}{n})$. (We're still going between some material and air.)

Note that total internal reflection can only occur when going from a material of higher index of refraction to a region of lower refraction.

From this brief description, can you guess what relative value of the index of refraction you'd want for optical fibers as opposed to the protective cladding on the fibers? An optical signal (light beam) injected into an optical fiber doesn't travel straight down the fiber from one end to the other. The signal bounces (reflects) off the walls of the fiber as it travels. You'd prefer to keep all of the signal contained. Hence, you want to maximize total internal reflection by making the fiber with a high index and the protective cladding with a low index of refraction.

---

[37] Warning! Keep all observations to under one minute.

# CHAPTER 6

# *Calculus*

*Take care of the pennies and the dollars will take care of themselves.* — **Scottish proverb**

## Attacking Calculus

Calculus is a practical tool. Remember, legend has it that Leibniz invented calculus to help design beer casks, and what is more practical or important than that? (Actually, both the English physicist and mathematician Sir Isaac Newton and the German mathematician Gottfried Wilhelm Leibniz are credited with inventing calculus.)

Calculus is also intuitively simple. Did we lose credibility with that statement? OK, we retract it for now, but bear with us.

Let's go back to functions which were introduced in an earlier chapter. A function takes a variable, operates on it, and gives you a value. That, of course, is very useful; we learn what is happening at any desired point. As we plot different points of the function we generate a curve[1]. With calculus we'll extract a lot more information out of that function. Using a few operations, we can learn what that function has been doing and where it's going.

With differential calculus we look at how a function changes shape as it meanders along our graph. We will take small (very, very small) differences in the independent variable and look at the changes in the value of our function. Does the value increase, decrease, or stay the same? The change in value of the function divided by the very small change in the independent variable is called the slope of the curve, and that is what differential calculus is all about.

---

[1] Our curve may well be a straight line, but we'll call it a curve anyway. A straight line is just a curve with an infinite radius of curvature.

By looking at the curve's slope, we might say that we are looking at where the curve is going. With integral calculus we look at the area under the curve. We again take a very small difference in our independent variable and look at the value generated by the function near that point. This time, however, we will multiply those two instead of dividing and get the area of that minuscule rectangle rather than the slope of the curve. Add a bunch of these mini areas together, and we have the area under the curve defined by our function. The neat little operation that we'll perform on our function to give us the area under its curve is what integral calculus is all about.

## Differential Calculus

Let's take some modern laboratory equipment and hop back in time to help Galileo measure the position of his cannonballs as they fall from that beautiful leaning tower. If we measure the ball's position every 0.2 second as it falls, we'll get data such as that plotted in Figure 6-1.

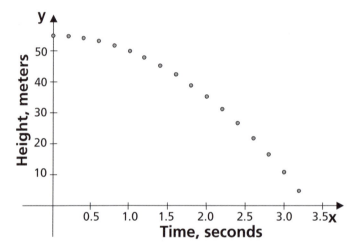

**Figure 6-1. Position vs. time of Galileo's cannon ball.**

What is the velocity of the falling ball? The whole 55-meter trip takes 3.35 seconds. Hence, the average velocity during the entire trip is 55 meters/3.35 seconds =16.4 m/s. By average velocity, of course, we mean the total distance traveled divided by the total time of travel. But this specific value of velocity is the actual value of velocity for only a brief moment during the fall. At the instant of dropping the ball, the velocity is zero, and just as the ball hits the ground it is really moving. If we want a better handle on the velocity as the ball falls, we can look at the distance covered during each 0.2 second and calculate the average velocity during each of these intervals. We get the results shown in Figure 6-2. By taking measurements over shorter intervals, we see that we are getting a better picture of the motion of the falling cannonball.

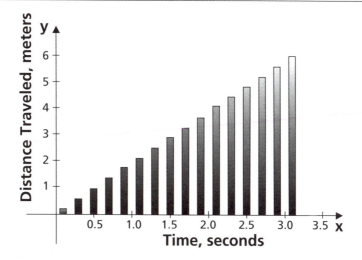

**Figure 6-2. Distance covered for each 0.2 seconds of fall.**

From the data shown in Figure 6-2 we can calculate the average velocity during any of those 0.2 second intervals as

$\bar{v}_i = \dfrac{d_i}{0.2\,\text{sec}}$ where the subscript, $i$, ranges from 1 to 16 since the cannonball hit the ground during the seventeenth time period, and $v_i$ is in meters/second.

These average velocities are plotted in Figure 6-3. We can see that the cannonball is going faster and faster during each succeeding time period, but wouldn't you like to be able to determine the velocity at any time during the fall?

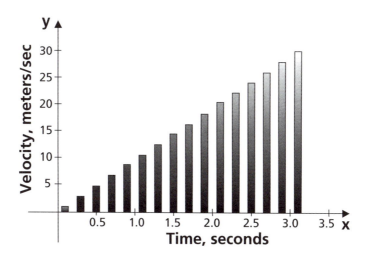

**Figure 6-3. Average velocity during each .02 seconds of fall.**

If we want information about the velocity at any chosen time, we need an expression that gives position as a function of time. Look at the measured data points and note that we can fit this data to the function:

$$y(t) = 4.9t^2.$$

(In later chapters we'll see how this is done. For now, simply note that the equation works as an expression of position as a function of time for this data.)

Now let's calculate the velocity at some arbitrary time that we'll call $t_1$. Of course, to calculate velocity from position data we need two different positions and the time difference between those two positions. So let's look at the position at $t_1$ and $t_1 + dt$, where $dt$ is a very small time interval.

So $\quad y(t_1) = 4.9t_1^2$

and

$$y(t_1 + dt) = 4.9(t_1 + dt)^2$$
$$= 4.9[t_1^2 + 2t_1 dt + (dt)^2]$$

and let's call this small change in position of the cannonball that occurs between the time intervals $dy$.

$$dy = y(t_1 + dt) - y(t_1)$$
$$= 4.9\{(t_1 + dt)^2 - t_1^2\}$$
$$= 4.9\{t_1^2 + 2t_1 dt + (dt)^2 - t_1^2\}$$
$$= 4.9\{2t_1 dt + (dt)^2\}$$

We went through a bunch of algebraic steps getting here. We'll not bore you so completely in future sequences, but we don't want anybody to get lost at this crucial stage.

This is where we'll make use of the fact that $dt$ is very, very, small, for if $dt$ is small—that is, much, much less than unity—then $(dt)^2$ gets a whole lot smaller than $dt$. Since $y(t)$ is a smooth, continuous function[2], we'll let $dt$ get so small that we can ignore $(dt)^2$. With $(dt)^2$ out of the way, the small change in position is simply

$$dy = 4.9 \bullet 2t_1 dt$$
$$= 9.8t_1 dt$$

What we have just completed is valid for any time, $t$, from drop to bounce; so, let's drop the subscript on $t$.

That done, we can now complete the simple function for the velocity at any time $t$.

---

[2] We'll discuss these terms later in this chapter. For now, just assume that our curves are neither broken nor jagged.

$$v(t) = \frac{dy}{dt}$$

$$= \frac{9.8t\,dt}{dt}$$

$$= 9.8t$$

(in units of meters/second)

which you may recognize as free-fall velocity, due to gravity, of an object near the earth's surface.

*Velocity* is the change in position of an object with respect to time. *Acceleration* is the change in velocity of that object with respect to time. We'll call acceleration *a(t)*. The velocity of the cannonball continues to increase at the same rate as time passes, so we have a constant acceleration. Let's calculate this acceleration.

The change in velocity between times $t$ and $t + dt$ is

$$dv = v(t + dt) - v(t)$$

$$= 9.8(t + dt - t) \quad .$$

$$= 9.8\,dt$$

So the acceleration of the cannonball is

$$a(t) = \frac{dv}{dt}$$

$$= \frac{9.8\,dt}{dt} .$$

$$= 9.8$$

(where *a(t)* has the units[3] meters/second$^2$)

---

[3]  Watch those units! Tracking the units during manipulations of different functions does not guarantee a correct answer, but if the units are wrong you are insured of a wrong answer. The position of the cannon ball is obviously length, for which we'll use meters. Time is in seconds, so velocity, length divided by time, is meters/second (m/s). Acceleration, velocity divided by time, is meters/second/second (m/s$^2$). "Where were these units hiding?" you ask. They reside with that multiplicative constant. When you measured the position of the falling ball you knew that its position had the unit of meters, but you also saw that the position changed as $t^2$. Hence to wind up with meters, that constant must have the units m/t$^2$. Remember, also, that subtraction does not change units, so $dt$ still has the unit of time.

Since time is no longer part of the function, we drop the *(t)* and say

$$a = 9.8\,\text{m/s}^2.$$

which is, of course, the acceleration near the Earth's surface due to gravity.

You have just taken the first (velocity) and second (acceleration) derivatives of the (position) function $y(t) = 4.9t^2$.

And now we define the *derivative* of the function y(t). It is

$$\frac{dy(t)}{dt} = \frac{y(t+dt) - y(t)}{dt}$$
**Remember this one!**  Eq. 6-1

as *dt* gets very, very small[4].

The position of the cannonball can now be plotted as a continuous function of time as shown in Figure 6-4.

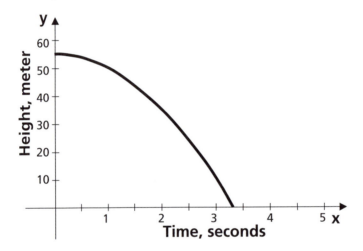

**Figure 6-4. Position of cannon ball as a continuous function of time.**

Let's move along and see if we can find a better way to calculate derivatives. I'd never keep track of the algebra for even the first derivative of a polynomial function like

$$y(x) = Ax^5 + Bx^4 + Cx^3 + Dx^2 + Ex + F$$

where *A...F* are constants, and we used the more conventional *x* as the variable instead of *t*.

---

[4]  Other books may use other symbols for the derivative such as $D_t y(t)$, $y'(t)$, or $\frac{\Delta y(t)}{\Delta t}$. Not to worry, they all mean the same thing.

Instead of jumping into a fifth-degree polynomial, let's go to one degree higher than what we've already done and look at

$$y(x) = Ax^3 + Bx^2 + Cx + D.$$

Proceeding as before,

$$dy = y(x + dx) - y(x)$$

$$= \{A(x + dx)^3 + B(x + dx)^2 + C(x + dx) + D\} - \{Ax^3 + Bx^2 + Cx + D\}$$

$$= \{A[x^3 + 3x^2(dx) + 3(dx^2)x + (dx)^3] + B[x^2 + 2x(dx) + (dx)^2] + C[x + dx] + D\} .$$
$$- \{Ax^3 + Bx^2 + Cx + D\}$$

Note that all of the terms associated with a given coefficient have the same degree of $x$. So combining terms with common coefficients we get

$$dy = A[3x^2(dx) + 3(dx)^2 x + (dx)^3] + B[2x(dx) + (dx)^2] + C[dx].$$

As before, we throw away any of those minuscule $dx$s that are squared or cubed or whatever else of higher degree and get

$$dy = 3Ax^2(dx) + 2Bx(dx) + Cdx.$$

The derivative is

$$\frac{dy}{dx} = \frac{3Ax^2(dx) + 2Bx(dx) + Cdx}{dx} .$$

$$= 3Ax^2 + 2Bx + C$$

We don't have a physical entity to relate to this equation, though if we were not so lazy we might find one. It's not needed since we're only looking for patterns.

Do you see the pattern? With polynomials you use the exponent as a new multiplier on each term and subtract one from that exponent. Note that each term in the polynomial is differentiated as though it stood alone. That is, the $B$ terms had no effect on the $A$ terms and so on. The constant, $D$ in the polynomial equation[5], disappears because a constant function has zero slope.

One very important thing to notice: no term in the polynomial affects any other term. As long as the function is a combination of terms added to each other, you just combine terms with common coefficients and degree of $x$ (note that $x^2 dx$, $x(dx)^2$, $(dx)^3$, and $x^3$ are all third degree in $x$).

---

[5] An additive constant in any function disappears upon differentiation for the same reason.

How about even higher degree polynomials? Well, we did a second degree and a second plus one—not quite a general $n$ and $n+1$ as a required proof via mathematical induction, but that's as far as we're going. We'll just assume that it works; too much algebra otherwise[6].

However, most functions are not nice neat polynomials. What do we do about trig functions, exponentials and such? Do we just throw in a $dx$ and crank away?

Yep, that's all we do. Let's look at a couple of them and then move along to something more exciting.

Since engineer/scientist types use lots of sines and cosines, let's look at them first.

We'll use $\theta$ as the variable because it's the convention, and let

$$y(\theta) = A \sin \theta .$$

We want to determine the derivative of $y(\theta)$ with respect to $\theta$. That is, we are looking for the slope of the curve described as $A \sin\theta$, for all values of $\theta$.

Using the same procedure as before, we start with the change in position, $y$, as before

$$dy = y(\theta + d\theta) - y(\theta)$$
$$= A \sin(\theta + d\theta) - A \sin \theta .$$

That $A \sin(\theta + d\theta)$ looks kind of nasty so let's see if it goes by an alias. Remember those trig identities we covered back in the trigonometry chapter?

Let's use the identity:

$$\sin(\theta + \alpha) = \sin \theta \cos \alpha + \cos \theta \sin \alpha$$

Let $\alpha = d\theta$

so that

$$dy = A \sin(\theta + d\theta) - A \sin \theta$$
$$= A\{\sin \theta \cos d\theta + \cos \theta \sin d\theta\} - \{A \sin \theta\}$$

---

[6]  I once had a polynomial equation with ten terms in it, and I needed its derivative (ref: *Journal of Crystal Growth*, vol. 147, pp 83-90, 1995). I had a hard enough time keeping track of coefficients and exponents without the threat of the algebra. To get the polynomial I used a computer program to find the coefficients to match a curve to measured data. I then performed a simple polynomial derivative to get what I needed.

This is still very messy until we make use of that wonderful property of $d\theta$. We let $d\theta$ get very, very small, and then see what happens.

Since we know from the trigonometry chapter that $\cos d\theta$ approaches unity as its argument, $d\theta$, approaches zero, we have

$$\sin\theta\cos d\theta \approx \sin\theta$$

as we shrink $d\theta$.

Likewise, $\sin d\theta$ approaches zero as its argument approaches zero, but here is the crucial difference. As $\sin d\theta$ approaches zero, it takes on the value of its argument, $d\theta$. Look at Table 6-1 if you don't believe us.

**Table 6-1. The value of the sine function as its argument approaches zero.**

| $\theta$ | $\sin\theta$ |
|---|---|
| $\pi/2$ | 1.0000 |
| 1 | 0.8415 |
| 0.1 | 0.0998 |
| 0.01 | 0.0100 |
| 0.001 | 0.0010 |
| 0.0001 | 0.0001 |

So, $\sin d\theta$ approaches $d\theta$ as $d\theta$ gets very, very small. Hence,

$$\cos\theta\sin d\theta \approx d\theta\cos\theta .$$

Now we're home free. With those substitutions (and we'll replace $\approx$ with $=$) we have

$$dy = A\{\sin\theta + d\theta\cos\theta\} - \{A\sin\theta\}$$
$$= A\cos\theta \cdot d\theta$$

Then

$$\frac{d\sin\theta}{d\theta} = A\cos\theta .$$  **Remember this one!** Eq. 6-2

A similar set of manipulations produces

$$\frac{d\cos\theta}{d\theta} = -A\sin\theta .$$  **Remember this one!** Eq. 6-3

We'll do one more single function and then move on to composite functions and compound functions.

Look now at

$\dfrac{dy(x)}{dx}$, where $y(x) = A^{ax}$.

Use the same procedure as before to determine $dy$.

$$dy = A^{a(x+dx)} - A^{ax}$$
$$= \{A^{ax} A^{dx}\} - A^{ax}$$

Any number (other than zero) approaches unity as its exponent approaches zero, but as with $\sin d\theta$, it's how that number approaches zero that's important. Let's look at some examples in Table 6-2.

**Table 6-2. Values of exponentials as the exponent approaches zero.**

|  | A = 3 | A = 2.71828 | A = 2.5 |
|---|---|---|---|
| $x$ | $3^x$ | $2.71828^x$ | $2.5^x$ |
| 1.00000 | 3.00000 | 2.71828 | 2.50000 |
| 0.10000 | 1.11612 | 1.10517 | 1.09596 |
| 0.01000 | 1.01105 | 1.01005 | 1.00921 |
| 0.00100 | 1.00110 | 1.00100 | 1.00092 |
| 0.00010 | 1.00011 | 1.00010 | 1.00009 |

Here we notice that when the number 2.71828 is raised to a very small exponent, the value rapidly approaches one plus that very small number. Some of you may recognize 2.71828 as an approximation for the base of the natural logarithm, funny little guy called[7] "*e*".

Now that we've argued that

$e^{dx} = 1 + dx$    as $dx$ becomes very small

then

if $y(x) = e^x$,

it follows that

---

[7] Other names in addition to *e* are *E*, *exp*, and base common log.

$$dy = e^{x+dx} - e^x$$

$$= e^x (1+dx) - e^x$$

$$= e^x \, dx$$

So

$$\frac{de^x}{dx} = \frac{e^x \, dx}{dx}.$$

$$= e^x$$

That is, the derivative of the function $e^x$ is $e^x$. This may not seem like a big deal, but this simple relationship becomes *very* important when we get into differential equations in a subsequent chapter.

Following the procedures that we've used before, you can derive other relationships involving our new friend, $e$.

Determine

$$\frac{df(x)}{dx} \text{ when } f(x) = e^{ax}.$$

$$\frac{de^{ax}}{dx} = \frac{\{e^{(ax+adx)}\} - e^{ax}}{dx}$$

$$= \frac{\{e^{ax}(1+adx)\} - e^{ax}}{dx}.$$

$$= ae^{ax}$$

and determine

$$\frac{df(x)}{dx} \text{ when } f(x) = e^{x^n}$$

$$\frac{de^{x^n}}{dx} = \frac{\left\{e^{\left(x^n + nx^{(n-1)}dx\right)}\right\} - e^{x^n}}{dx}$$

$$= \frac{e^{x^n}(1 + nx^{n-1}dx) - e^{x^n}}{dx}$$

$$= nx^{n-1}e^{x^n}$$

As before, we dropped the higher-degree terms of $(dx)$ when we expanded $(x+dx)^n$.

That last function is an example of a *composite function*, a function where the argument is also a function of the independent variable, $x$. Let's try that one again by using an example.

Let $f(t) = \sin t$

and $t = x^2$

so the function, $f(t)$, is really a function of $x$.

We could get fancy and write it as

$$f(t) = f\{t(x)\}$$

$$= f(x^2)$$

$$= \sin(x^2)$$

Now, let's differentiate that guy and see how it changes with respect to $x$.

$$\frac{df(t)}{dx} = \frac{df[t(x)]}{dx}$$

$$= \frac{d\sin(x^2)}{dx}$$

$$= \frac{\sin[(x+dx)^2] - \sin(x^2)}{dx}$$

$$= \frac{\sin(x^2 + 2xdx) - \sin(x^2)}{dx}$$

$$= 2x\cos(x^2)$$

Notice that we used our friendly trig identities[8] in going from step 4 to 5 in the last equation, and that we dropped terms with $dx$ to powers greater than one such as $(dx)^2$. Also, please remember that $\sin^2 x \neq \sin(x^2)$.

Looking again at the form of the *f(t)* function,

$$f(t) = f[t(x)]$$

see that we used what is commonly called the *Chain Rule* to determine the derivative of $\sin(x^2)$.

---

[8] We used the identity and property, $\sin(a+b) = \sin a \cdot \cos b + \cos a \cdot \sin b \approx \sin a + b \cos a$, as $b$ gets very small.

In symbols, the Chain Rule is as follows:

$$\frac{df[t(x)]}{dx} = \frac{df[t(x)]}{d[t(x)]} \cdot \frac{dt(x)}{dx}$$

**Remember this form.** Eq. 6-4

It's simple. To take the derivative of a function where the argument is another function of the independent variable, you just take the derivative of the outside function as though the argument was just the simple variable[9]. Then you multiply this result by the derivative of the inside function. We just showed the Chain Rule with an example.

## Derivatives of Compound Functions

Derivatives of compound functions are easy after all that you have mastered.

A compound function is a function that can be written as

$$y(t) = f(t) \cdot g(t).$$

As an example, we'll pick

$$y(t) = At^2 \sin t$$

where

$$f(t) = At^2 \quad \text{and}$$

$$g(t) = \sin t.$$

To differentiate $y(t)$ we'll crank as before and get

$$\frac{dy(t)}{dt} = \frac{d\{At^2 \sin t\}}{dt}$$

$$= \frac{A\{(t+dt)^2 \sin(t+dt)\} - A\{t^2 \sin t\}}{dt}$$

$$= \frac{A\{(t^2 + 2tdt)\sin(t+dt)\} - \{At^2 \sin t\}}{dt}$$

$$= \frac{A\{(t^2 + 2tdt) \cdot (\sin t \cos dt + \cos t \sin dt)\} - \{At^2 \sin t\}}{dt}$$

$$= 2At \sin t + At^2 \cos t$$

---

[9] Or in simpler words, the Chain Rule says to *take the derivative of the outside and multiply it by the derivative of the inside.*

Examine the terms, and you'll see that the derivative of that compound function is the derivative of the first term times the second term plus the first term times the derivative of the second term.

That is,

$$\frac{dy(t)}{dt} = \frac{d}{dt}[f(t) \cdot g(t)]$$

$$= \frac{df(t)}{dt} \cdot g(t) + f(t) \cdot \frac{dg(t)}{dt}$$

**Remember this one!**   Eq. 6-5

Will this always work? Try it on a few more examples and see for yourself.

# The Importance of Zero Values of Derivatives

We'll now divert from pure math into physics, engineering, and economics with examples to convince any doubters among you that we really have a great tool here.

For starters, let's assume that Galileo's assistant got playful one day and threw one of the lighter cannon balls upward with an initial velocity of –10 m/s. (For simplicity, we are taking the top of the bell tower as $y = 0$ and letting $y$ increase (positive) towards the ground.)  The ball, of course, starts going up as gravity is trying to slow it and reverse its direction. Let's further assume that Galileo is intrigued by this new experiment and with careful observations fits this data to the following equation:

$$y(t) = V_0 t + \tfrac{1}{2} A t^2$$

where:

$V_0$ = the initial velocity of the thrown cannonball (it will take a negative value as it is thrown upward)

$A = 9.8$ m/s$^2$, the acceleration due to gravity, as we determined earlier.

If you should want to know how high you can expect the ball to go for any initial upward push, $V_o$, you wish to give it, just apply differential calculus. Let's think about how to proceed. The ball starts upward at some velocity. Sometime later it hits the ground with another velocity in the opposite direction. Somewhere in between the ball had to stop and change direction, and that, of course, would be at its maximum height (in the negative $y$ direction). So, the maximum height coincides with the change in direction of the ball, which occurs when its velocity equals zero.

Where does calculus come into play? You've already seen that the first derivative of position with respect to time is velocity, so take the derivative of this new function and you get

$$v(t) = \frac{y(t)}{dt}$$

$$= \frac{d}{dt}[V_0 t + \tfrac{1}{2}At^2]$$

$$= V_0 + At$$

To determine when the velocity equals zero, look at the algebra of our recently differentiated equation. That is,

$v(t_m) = 0$, when $V_0 + at_m = 0$, where $t_m$ = the time when the cannonball reaches its maximum height above the tower and then starts back down.

Hence,

$$t_m = -\frac{V_0}{A} \,.$$

That is, the ball stops climbing in $\frac{-V_0}{A}$ seconds and then starts falling toward the ground. So if $V_0 = -10$ m/s, then the turn-around occurs approximately 1 second after the toss. At the turnaround point, the velocity is equal to zero for a brief moment of time.

"How high did it go?" you ask. To determine the height, just plug $t_m = 1$ s into your height equation and get a height above the tower of 10 meters.

The position (height) of the two cannonballs as a function of time in shown in Figure 6-5.

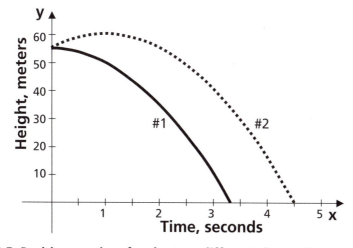

**Figure 6-5. Position vs. time for the two different descending cannon balls.**

And here is the real lesson of this little section: Setting the first derivative to zero determines a relative maximum or minimum of the function. What do we mean by *relative*? Let's look at another example.

Look at the function

$$f(x) = 10x - 2x^2 + 0.1x^3 - 0.0014x^4 .$$

Since this particular function has no physical significance that we know of, let's assume it describes the motion of a race car needing a tune-up.

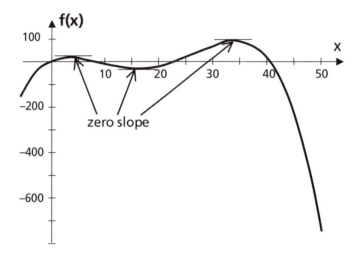

**Figure 6-6. Plot of polynomial with two maxima and one minimum.**

In Figure 6-6, *f(x)* has two peaks and one valley. There are two simple but important concepts here: 1) the first peak for this function is a relative maximum because there is another larger one coming along, and 2) once you reach a relative maximum, you must go through a relative minimum before you can reach another relative maximum.

And how do we determine where these peaks and valleys (relative maximum and minimum) occur? Since it is obvious the slope of the plotted function (the slope of the curve) must be flat as the curve goes through these maxima and minima, and a flat spot on a curve is tantamount to a zero slope, then the first derivative must be zero at these peaks and valleys. So we determine these interesting points by looking at where the first derivative equals zero. That is, when

$$\frac{df(x)}{dx} = 10 - 4x + 0.3x^2 - 0.0056x^3 = 0$$

which occurs when

$x = 3.24, 16.11,$ and $34.23$ (within round-off errors).

Setting the value of the first derivative to zero tells us that the curve has reached a local maximum or minimum value, but we are not told which it is. To determine whether we have found a maximum or minimum, look at the second derivative of the function. Of course you could have just plotted the function and looked at the curve, but let's do it mathematically.

The first derivative (slope of the original curve) is positive (the function is increasing in value) as it approaches a local maximum. It is zero at that maximum, and then it flips over and becomes negative (the function is now decreasing in value) past that peak. See Figure 6-7. That much is fairly intuitive. Now consider the second derivative. The second derivative, which is the slope of the curve created by plotting the first derivative must be negative as the original function reaches a local maximum. This may not be obvious, but consider that even though the function was increasing before the maximum, it wasn't increasing as fast as its previous interval. The first derivative had to slow down for the original function to reach a plateau, hence the first derivative was decreasing as the initial curve reaches a maximum point. If the first derivative is decreasing, then the second derivative is negative. The second derivative is plotted in Figure 6-8. The reverse argument holds in the region of a local minimum. To visualize this concept, just consider a straight-edge ruler swinging around these peaks.

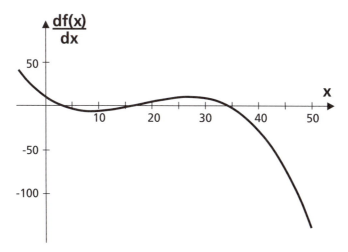

**Figure 6-7. First derivative of the curve in Figure 6-6.**

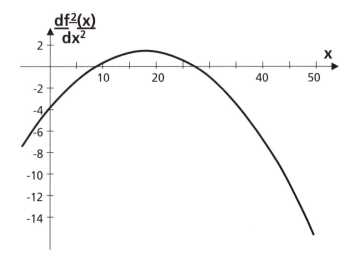

**Figure 6-8. Second derivative of the curve in Figure 6-6.**

So let's look at the second derivative of our function, which is typically written as

$$\frac{d}{dx}\left\{\frac{df(x)}{dx}\right\} = \frac{d^2 f(x)}{dx^2} \, .$$

$$\frac{d^2 f(x)}{dx^2} = -4 + 0.6x - 0.0168x^2$$

This equals –2.23, +1.30, and –3.14 at the three values of $x$ where the first derivative equals zero. Since our function was nicely plotted out in Figure 6-6, we can use eyeball determination to pick maxima and minima, but it's comforting to see that the signs of these derivatives follow the argument calling for negative values of the second derivative at local maxima and positive values at local minima.

"What, if anything," you might ask. "is the significance of where this second derivative equals zero?" (It has roots at $x = 8.9$ and 26.8.) Since the second derivative is negative at any given maximum and positive at the next local minimum, then it must be zero somewhere in between. Obvious, huh? We hope it's also obvious that where the second derivative equals zero is where the first derivative stops decreasing and starts to increase once again. That is really the point where the curve starts to turn around. These points where the second derivative equals zero are called the *inflection points* of the curve. The same argument with opposite signs holds if you were going from a local minima to the next maxima.

Inflection points and maximums and minimums are not reciprocal. If you have a maximum and a minimum, you will have an inflection point. The converse, however, is not necessarily true. Curves exist that have an inflection point but have no maximum or minimum. Take the function:

$$f(x) = 0.2x^3 - .02x^2 + 15x \quad \text{plotted in Figure 6-9.}$$

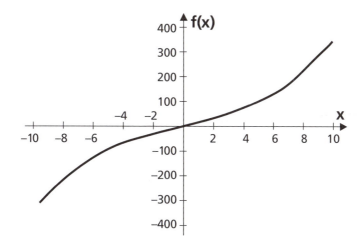

**Figure 6-9. Plot of a curve with an inflection point but no maximum or minimum. The second derivative of the curve *f(x)* = 0.2*x*³ − 0.02*x*² +15*x* equals zero at *x* = 3.3.**

If you take the first derivative and try to set it to zero, you'll see that it has no real roots. There is no maximum or minimum. But the second derivative is readily set to zero at $x = 0.033$. Looking at the plot, you can see that the slope stops decreasing and begins to increase at that value even though no maximum or minimum occurred.

Another way to look at the inflection point is that it is the value of the independent variable where the tangent to the curve is under the curve on one side of that point (the inflection point) and on top of the curve on the other side. Take your straight-edge and check this out.

Let us bore you with one last property of the inflection point, and then we'll move along to other good stuff. The inflection point tells you when the curve goes from being convex to concave or vice versa. Of course, whether it's convex or concave depends on whether you're looking up or down. Let's not get tangled in semantics—just look at an example and check it out.

## Other Topics

We have left out some important topics about which the mathematical purists will scold us, so let us at least approach them in this mix of items.

## Topic #1. The Limit

"Purer" math books will define the derivative as

$$\frac{df(x)}{dx} = \lim_{dx \to 0} \frac{f(x+dx) - f(x)}{dx}.$$ which to us, is the same as saying $dx$ gets very, very small.

Discussions of the limit are about *how small is small enough* when you want to eliminate the $(dx)^2$ and higher-degree terms, and when we let $\sin(dx) = dx$, and $e^{dx} = 1 + dx$, and other such minutia (pun intended). These discussions are quite important to some people—whole books have been written on this subject—but that's not what this book is about.

## Topic #2. Some Special Functions and Some Special Considerations

If we consider two interesting functions

a)   $y(x) = |x|$, that is, $y(x)$ = the absolute value of $x$

and

b)   $g(x) = x - [x]$, where $[x]$ is defined as the greatest integer equal to or less than $x$

we can discuss just where the derivative has meaning.

The first function, $y(x) = |x|$, has valid properties and is continuous[10] over the entire domain of finite numbers. This function has an easily calculated derivative at all values of $x$ except for one. How do we calculate the derivative at $x = 0$? Simple answer: we can't. Once past zero on the positive $x$ side, the derivative is +1. Once past zero going negative, the derivative is –1. At zero we encounter a sharp break, and the derivative has no meaning at a juncture such as this. So this function is *continuous*, but it is not *smooth*.

The second function, $g(x) = x - [x]$, is even worse. Even though functions that look like this one are used to pull the electron beam across your computer screen, you can't readily handle them with ordinary math[11] for it is not even continuous. It has no unique value when $x$ equals an integer. And without a unique value, we can't even start to think about a derivative in its normal sense. This function is neither continuous nor smooth at $x$ = an integer.

---

[10] The curve has no gaps in it. That is, for every $x$ there is one and only one value for $f(x)$.

[11] Note to the electrical engineers in the audience: Oliver Heaviside developed a set of functions that bear his name for just such applications as the circuits that control your computer screen.

## Topic #3. Interval of Existence

Often we find a function that describes our physical world over some range of values of the independent variable, but then the function becomes meaningless as that independent variable takes on other values. Look at Galileo's falling cannon ball. We fit a nice function for the ball's position during the 3.3-second fall, but we have no function to describe the ball's trip up the stairs, although we can sympathize with the graduate student's aching back during that 294-step climb. Although physics may control the ball's motion after it hits the ground, we don't want to try to formulate a function for its motion as it bounces about. (Among the variables to consider after Earth contact are the softness of the ground—has it rained lately—the slope of the ground, and the number of graduate students trying to catch the ball.)

Our stated function only exists for the time during the actual fall—that seems obvious enough. However, if we look at the two functions in the last section, $y(x) = |x|$, and $g(x) = x - [x]$, it is also obvious that interval of existence for the derivative exists for all values other than $x = 0$ for $y(x) = |x|$, though the function exists for all values of $x$. And in the second function, neither the function nor its derivative exists for $x$ equal to an integer.

## Topic #4. Series

You will run into many different types of series as you bop along in the mathematical and engineering world, but we are only going to touch on the subject lightly. Most of what we'll discuss will seem obvious to you, but then again, doesn't most of this stuff seem simple after going over it?

If we take a function such as $f(x) = \dfrac{1}{x}$, where $x$ is an integer, then $f(x)$ is a *sequence* and may be displayed[12] as $1, \dfrac{1}{2}, \dfrac{1}{3}, \dfrac{1}{4}, \cdots, \dfrac{1}{x}, \cdots$. If we add the terms of a sequence together, we have a *series*. So $f(x) = \dfrac{1}{x}$ is a sequence, and $\sum_i f(x_i)$ is a series. If the summation goes on without end, we have an infinite series. If the series sums up to some finite value, then we say that we have a convergent series, and the series converges to that value. Obviously the terms must become vanishingly small as the series becomes large if it is to converge.

---

[12] Be careful in going the other way. That is, if you looked at only the first three terms of this sequence you might have assumed the generating function was $f(x) = \dfrac{1}{x}$ as shown, but if the fourth term was $\frac{3}{4}$ then the generating function would have been $f(x) = \dfrac{x + (-1)^x (x-2)}{2x}$. Check it for yourself and see that the first three terms of these two functions are the same.

We will use a series of simple functions to represent complicated functions. A good example of this is the Fourier series of sines and cosines which you'll see later in this chapter. Numerical analysis techniques use copious series solutions to help the computer crank away on your engineering problems. You will also use Taylor's and Maclaurin's expansions (which are really series) to represent functions.

### Taylor's and Maclaurin's Expansions[13]

These expansions or approximations rest on the idea that the value of $f(x)$ at $x=b$ can be estimated if we know the value of $f(a)$, where $a \neq b$, and if we can differentiate $f(x)$. That is,

$$f(b) = f(a) + \frac{df(a)}{dx}(b-a) + \frac{d^2 f(a)}{dx^2}\frac{(b-a)^2}{2} + \frac{d^3 f(a)}{dx^3}\frac{(b-a)^3}{2*3} +$$
$$+ \frac{d^{(n-1)}f(a)}{dn^{(n-1)}}\frac{(b-a)^{n-1}}{(n-1)!} + \frac{d^n f(\xi)}{dx^n}\frac{(b-a)^n}{n!}$$

where $\frac{d^i f(a)}{dx^i}$ is the $i^{th}$ derivative of $f(x)$ evaluated at $x=a$, $n!$ is shorthand for n-factorial[14], and $\xi$ is a number between $a$ and $b$. The terms in this expansion must become smaller as their number increases if the series is going to be useful.

The validity of this expansion is obvious for linear equations. Let's try it for $f(x)=e^{-x}$ where $a=0$ and $b=1$. We know that $\frac{de^{-1}}{dx}=-1$ when $x=0$, so we'll proceed as

$$f(1) = e^{-1}$$
$$\approx 1 - 1 + \frac{1}{2} - \frac{1}{6} + \frac{1}{24}$$
$$\approx 0.375$$

which is within 2% of the actual value of $e^{-1}$. If we had taken more terms we would have gotten a better answer.

For a final example, let's use Maclaurin's expansion to represent $\sin\theta$ as a series[15].

$$\sin\theta = \sin(0) + \frac{d\sin(0)}{d\theta}(\theta) + \frac{d^2\sin(0)}{d\theta^2}\frac{(\theta)^2}{2} + \frac{d^3\sin(0)}{d\theta^3}\frac{(\theta)^3}{2*3} + \cdots + \frac{d^{(n-1)}\sin(0)}{d\theta^{(n-1)}}\frac{(\theta)^{n-1}}{(n-1)!} + \cdots.$$

---

[13] Maclaurin's expansion is as shown with the provision that $a=0$. If $a \neq 0$, then it belongs to Taylor.
[14] The term $n! = 1*2*3*\cdots(n-1)*n$. For example $4! = 1*2*3*4 = 24$.
[15] Series where the argument of the function takes on succeedingly higher orders are called *power series*.

As we know, $\sin(0) = 0$, $\dfrac{d\sin(0)}{d\theta} = \cos(0) = 1$, and so on. Therefore,

$$\sin\theta = 0 + \theta + 0 - \frac{\theta^3}{2*3} - 0 + \frac{\theta^5}{2*3*5} - \cdots + \frac{d^{(n-1)}\sin(0)}{d\theta^{(n-1)}}\frac{(\theta)^{n-1}}{(n-1)!} + \cdots$$

$$= \theta - \frac{\theta^3}{2*3} + \frac{\theta^5}{2*3*5} - \cdots + \frac{(-1)^n}{(2n+1)!}\theta^{2n+1} + \cdots$$

You'll use this series in the chapter on differential equations.

## Topic #5. L'Hôpital's Rule[16]

If you have two functions, $f(x)$ and $g(x)$, that are both equal to zero at the same value of $x$, you will not be overly concerned unless you are dividing one of the functions by the other one. If that is the case, your first thought might be to scrap the whole endeavor and go for coffee. L'Hôpital didn't drink a lot of coffee, so he came up with a very useful rule which is:

Given a function expressed as $\dfrac{f(x)}{g(x)}$ where $f(x)$ and $g(x)$ each approach zero

as $x$ approaches some number $c$, then $\dfrac{f(x)}{g(x)} = \dfrac{df(x)/dx}{dg(x)/dx}$ as $x$ approaches $c$.

An example is to calculate

$\dfrac{x^2}{x}$ as $x$ approaches zero.

Taking the derivatives of the top and the bottom[17], we get

$$\frac{x^2}{x} = \frac{2x}{1} = 0 \cdot$$

If we had tried to calculate $\dfrac{x}{x^2}$ we would have wound up with

$$\frac{x}{x^2} = \frac{1}{2x} = ?$$

We use ? instead of the infinity sign that we said we wouldn't use.

---

[16] Actually L'Hôpital had many rules. We are only going to talk about one of them, and even then we are not going to prove it.

[17] Be warned! You take the derivatives of the numerator and the denominator separately. You do not differentiate the fraction as a compound differentiation. And warning number two: do not use L'Hôpital's rule unless both functions approach zero at the same value of $x$.

What if we had a function such as $\dfrac{x^3}{x^2}$ and wanted to evaluate it at $x = 0$? We see that the first differentiation leaves us with something that we still can't use, so we differentiate again. That is, as $x$ approaches zero,

$$\frac{x^3}{x^2} = \frac{3x^2}{2x}$$
$$= \frac{6x}{2} \cdot$$
$$= 0$$

The foregoing examples were a bit simple. All of the interesting $x$ values were zero and all of the functions wound up equaling zero or some unmentionable number. It doesn't have to be this way. Look at $\dfrac{e^{x-1} - e^{1-x}}{\sin(x-1)}$ as $x$ approaches 1.

$$\frac{e^{x-1} - e^{1-x}}{\sin(x-1)} = \frac{e^{x-1} + e^{1-x}}{\cos(x-1)}$$
$$= \frac{e^0 + e^0}{\cos(0)} \cdot$$
$$= 2$$

Before moving on to integral calculus, note that a table of commonly used derivatives can be found in Appendix A.

## Integral Calculus

Consider some simple function such as

$$f(x) = x.$$

Rather than ask for the slope of its curve at any arbitrary value of $x$ (which you can now easily calculate by taking the derivative of the function), we now ask what other function would produce this $f(x)$ had we taken its derivative. In other words, what is the $F(x)$ such that

$$\frac{dF(x)}{dx} = f(x)$$

or what function, when differentiated, would equal $f(x)$?

We hope that all of you followed the earlier calculus section sufficiently to be able to work backwards and figure out[18] that $F(x) = \frac{1}{2}x^2$. For the time being, let's call this the antiderivative.

That was easy enough, and we bet you can determine what functions would produce a variety of derivatives such as sines, exponentials, and all sorts of polynomials. Once you think you have figured the source function of your derivative of interest, you can check your answer by performing that differentiation.

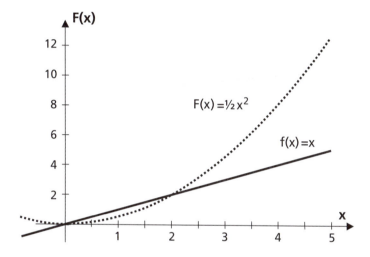

**Figure 6-10. Plot of the function *f(x)* = x and its antiderivative.**

When you take the derivative of our simple function you determine the slope of said function. What do you suppose you get when you determine the antiderivative of your function? Let's plot both of them and see what they look like.

Not very exciting, but look carefully and you'll see that, for any position along the x axis, the value of $F(x)$ at that position gives the area under the curve of $f(x)$ from the $x$ origin (i.e., from $x = 0$) to that chosen value of $x$.

---

[18] Our $F(x) = \frac{1}{2}x^2$ is not a unique antiderivative of $f(x) = x$, because another solution could just as easily be $F(x) = \frac{1}{2}x^2 + 34$. That is, any constant added to our simplest $F(x)$ is also an antiderivative because the constant goes away when we differentiate.

As you observed, the simple function we chose, $f(x) = x$, describes a triangle. You learned in grade school that the area under a triangle is equal to 1/2*base*height, which in this case is $\frac{1}{2}x^2$. Have we come upon something significant in discovering that the antiderivative can calculate the area under a curve? Or, have we only stumbled onto an example of simple geometry? Let's hope it is the former, because if it is, we can then determine areas under and between all types of curves, which will help explain all sorts of phenomena.

To show the equivalence of the antiderivative and the area under the curve generated by this function, let's first look at how we might calculate, at least conceptually, the values of these areas if we didn't know the equation for the area of that particular geometry. (For example, if our function was $f(x) = 3x^3 + 4x^2 + \sin x$, we wouldn't know how to use simple geometry to get an analytical expression for the area under the function between $x = 14$ and $x = 33$. Let's hope that we can come up with something using calculus.)

Any two-dimensional rectangular area in Cartesian coordinates is simply calculated as the product of base times height. In the case of the triangle generated by $f(x) = x$, we used the average height, $\frac{1}{2}x$, times the base, $x$. If we are dealing with any smooth, continuous curve and only look at the area under the function for a very small change in $x$, then the function will change only by a minute amount from one side of that interval to the other side, and the area of that small section can be approximated by a rectangle. The smaller the interval in $x$, the closer this approximation becomes an equality, and we have an exact value. (Have you heard that argument before?)

Who cares about tiny areas of such small intervals? Nobody! The concept, however, is useful. Remember the old Scottish proverb that said, "Take care of the pennies and the dollars will take care of themselves." So let's take a bunch of these little areas, add them up, and see what we get.

Let's look at any well-behaved arbitrary function, $g(x)$, and calculate the area under its curve between any two points on the independent variable axis, such as over the interval $x = a$ to $x = b$.

Our first step is to break that interval into very small but equal steps we'll call $\Delta x$, where $\Delta x = \dfrac{(b-a)}{N}$, and $N$ is a very large number, which makes $\Delta x$ very small. We now have a series of very small segments, and we can estimate the area of each of them from the product of $\Delta x$, the length of the segment, and $g(x_n)$, the approximate height of that segment where $a \leq x_n \leq b$. Of course, the height changes from one side of the segment to the other, but as the segments become smaller and smaller, the closer $g(x_n)$ approaches a single value. Your *leap of faith* here is to believe that the approximation becomes exact as $\Delta x$ becomes ever smaller.

That done, we now add up the small areas and, *voila*, we have the area under the curve. Let's put that in the form of an equation[19].

$$Area = \sum_{n=0}^{N} \{g(a + n \cdot \Delta x)\} \cdot \Delta x, \text{ where } \Delta x = \frac{(b-a)}{N}.$$

This is not very useful in this form, but we hope you follow the concept.

## The Fundamental Theorem of Calculus

The fundamental theorem of calculus shows the relationship between differential calculus and integral calculus, but it's a bit messy. If you want to skip this section and go to Results of the Fundamental Theorem of Calculus, we'll understand. But if you do skip it, please come back later.

Let's do another summation. In fact, let's do two summations. Let's sum $g(t)$ (yes, we changed the symbol of the argument, and you'll see why shortly) over the slightly different interval of $t = a$ to $t = x$, and we'll call that summation $G(x)$. And we'll do another summation of $g(t)$ and go from $t = a$ to $t = (x + h)$ and call it $G(x+h)$.

$$G(x) = \sum_{n=0}^{N} \{g(a + n \cdot \Delta t)\} \cdot \Delta t, \text{ where } \Delta t = \frac{(x-a)}{N}$$

and

$$G(x + h) = G(x) + \sum_{k=0}^{K} \{g(x + k \cdot \frac{h}{K})\} \cdot \frac{h}{K}$$

$$= \sum_{n=0}^{N} \{g(a + n \cdot \Delta t)\} \cdot \Delta t + \sum_{k=0}^{K} \{g(x + k \cdot \frac{h}{K})\} \cdot \frac{h}{K}$$

Such that

$$G(x + h) - G(x) = \sum_{k=0}^{K} \{g(x + k \cdot \frac{h}{K})\} \cdot \frac{h}{K}$$

Let's divide both sides of the last equation by $h$ to get

$$\frac{G(x + h) - G(x)}{h} = \frac{1}{K} \sum_{k=0}^{K} g(x + k \cdot \frac{h}{K}).$$

We hope the form of the left-hand side brings a bit of *deja vu* for you.

---

[19] If you want to get picky with us you might notice that we divided our interval, *a* to *b*, into *N* segments, and the summation from 0 to *N* is actually *N*+1 segments. If you want to be precise, we added an extra segment. But remember, these are vanishingly small intervals, and the bookkeeping is much simpler, as you will see, if we start the count at 0 instead of 1.

Somewhere over the $x$ and $x + h$ interval, $g(t)$ will have a maximum value and a minimum value[20], such that

$$g(t_{min}) \leq g(x + \frac{kh}{K}) \leq g(t_{max}), \text{ for all values of } k.$$

If we take all values of the dependent variable over the interval, we'll have

$$K \cdot g(t_{min}) \leq \sum_{0}^{K} \{g(x + \frac{kh}{K})\} \leq K \cdot g(t_{max}).$$

Divide through by $K$, and we have the right-hand side of the equation

$$\frac{G(x+h) - G(x)}{h} = \frac{1}{K} \sum_{n=0}^{N} g(x + k \cdot \frac{h}{K})$$

between $g(t_{min})$ and $g(t_{max})$.

That is

$$g(t_{min}) \leq \frac{G(x+h) - G(x)}{h} \leq g(t_{max}).$$

Let's think small once more and let $h$ get very, very small—so small that $g(t_{min})$ approaches $g(t_{max})$, and since the lower bound of that interval is $x$ then these $g(t)$'s approach $g(x)$. Hence, the inequalities in the last equation go to equal signs. Or

$$g(x) = \frac{G(x+h) - G(x)}{h}.$$

and that gives us a hand-waving proof of the Fundamental Theorem of Calculus, which is

$$g(x) = \frac{dG(x)}{dx},$$

or in words:

## The Results of the Fundamental Theorem of Calculus

The area under the curve described by $g(x)$ over the interval $a$ to $x$ is equal to $G(x)$, where $G(x)$ is the antiderivative of $g(x)$.

Or in words, the antiderivative of a function gives us a method to determine the area under the curve generated by that function.

That's enough of proofs for a while. Let's use what we've learned.

---

[20] We are not specifying that a zero value of the first derivative must accomplish this maximum and minimum. Even if $g(t)$ = constant over this interval, then max = min, and we keep on chugging.

First, let's simplify the notation. We can do away with the summation notation since the antiderivative calculates the area under the curve. Of course, we may not always want the area from the graph origin to whatever end point we choose. So if we want the area under the curve, $g(x)$, between points a and b, we just determine the antiderivative, $G(x)$, and take the difference of the value of this function at the two values of $x$.

That is, the area under $g(x)$ between points $a$ and $b$ (where a<b) is $G(b)-G(a)$, which we'll write as

$$\int_a^b g(x)dx = G(b)-G(a).$$

Let's stop using the term *antiderivative* and use a more sophisticated term. We'll use *integral*, when we speak of finding the function that computes the area under the curve. If we only refer to the function itself without determining the area under a specified part, then it's an *indefinite integral*. Once we put values on it, we have a *definite integral*.

The indefinite integral of $f(x) = x$ is $\frac{1}{2}x^2$, and the definite integral of this

$f(x)$ over the range $3 \le x \le 8$ is equal to 27.5. Or in math terms

$$\int xdx = \frac{1}{2}x^2$$

and

$$\int_3^8 xdx = \frac{1}{2}\cdot 8^2 - \frac{1}{2}\cdot 3^2$$
$$= 27.5$$

Typically, the indefinite and definite terms are left off, and we refer to the function or the value as the *integral*, and the process is called *integration*. If a range is specified, it's a definite integral; otherwise, it's an indefinite integral. Just to include everybody, we call the function $f(x)$ in this example the *integrand*.

Let's look at some basic formulas in integration.

1) $\int \{c \cdot f(x)\}dx = c \cdot \int f(x)dx$, where c is a constant

2) $\int \{f(x) \pm g(x)\}dx = \int f(x)dx \pm \int g(x)dx$

3) The integral

$$\int x^n dx$$

has to be written in two parts. It is

$$\int x^n dx = \frac{1}{n+1} x^{n+1} \text{ , if } n = -1. \text{ However if } n = -1, \text{ then}$$

$$\int x^{-1} dx = \ln|x| .$$

4)   $\int \{df(x)/dx\} dx = f(x) + C$  , where $C$ is a constant.

These look easy but don't go too far and say that if you can differentiate then you can integrate. That may be true in theory, but this is a practical book. It's easy enough to differentiate $F(x) = \sin^3 x$  and get $f(x) = 3 \cdot \sin^2 x \cdot \cos x$ . You just perform the differentiation of a composite function as you learned earlier. However, we would never casually look at $3 \sin^2 x \cos x$  and readily determine that its integral is $\sin^3 x$ . Help is available. Some nice people over the years have looked at all sorts of functions and put together extensive tables[21] of integrals. We've even included a table of common integrals in Appendix A.

Sometimes integration gets even messier.  The technique we'll now discuss is called *integration by parts*. It's a bit tricky so bear with us; it's worth it.

Assume that you're having a tough time trying to integrate a function. You look carefully at that function and see that it can be written as a product of two functions.

That is, you're trying to integrate some $f(x)$ . You're having trouble getting started and can't even find the solution in the integral tables. Then you look at $f(x)$  and see that it can be written as

$$f(x) = u(x) \cdot \frac{dv(x)}{dx} .$$

That is, $f(x)$  can be written as the product of the function $u(x)$  and as the derivative of the function $v(x)$ .

Then, the integral would be

$$\int f(x) dx = \int \{u(x) \cdot \frac{dv(x)}{dx}\} dx$$

Writing part of the product as the derivative of another function looks awfully messy, but there's a method in this.

We know that

$$\int \{\frac{d}{dx}[u(x) \cdot v(x)]\} dx = u(x) \cdot v(x) .$$

---

[21] *CRC Standard Mathematical Tables*, 27th Edition, W. H. Beyer, ed, CRC Press, Boca Raton, FL, 1985.

But the differentiation of a product of two functions, as you learned earlier, looks like

$$\frac{d}{dx}\{u(x) \cdot v(x)\} = u(x) \cdot \frac{dv(x)}{dx} + \frac{du(x)}{dx} \cdot v(x)$$

Thus, substituting the right-hand side from above for our integrand, we get

$$\int \{(u(x) \cdot \frac{dv(x)}{dx} + \frac{du(x)}{dx} \cdot v(x)\}dx = u(x) \cdot v(x)$$

then,

$$\int \{u(x) \cdot \frac{dv(x)}{dx}\}dx + \int \{\frac{du(x)}{dx} \cdot v(x)\}dx = u(x) \cdot v(x)$$

After rearranging, we can write our original formula as

$$\int f(x)dx = \int \{u(x) \cdot \frac{dv(x)}{dx}\}dx$$

$$= u(x) \cdot v(x) - \int \{\frac{du(x)}{dx} \cdot v(x)\}dx \qquad \textbf{\textit{Remember this one!}} \quad \text{Eq. 6-6}$$

The trick now is to determine how to break up our $f(x)$ and choose which part of the product to call the derivative. Poor selection of the $u$ and $v$ can really complicate the situation. It's usually better to start by choosing, for $u(x)$, the part of the product that makes $\frac{du(x)}{dx}$ simpler than $u(x)$.

As an example, look at

$$\int \{x \cdot \cos x\}dx \text{ and try to integrate it.}$$

If your integral table book lists it, then you're home free. Otherwise, you can try to put it into the form to solve by integration by parts.

If we ignore the advice on the choice of $u(x)$ given above and choose

$$u(x) = \cos x, \text{ and } \frac{dv(x)}{dx} = x, \text{ which makes } \frac{du(x)}{dx} = -\sin x \text{ and } v(x) = \frac{1}{2}x^2,$$

and then by plugging into the integration by parts equation, we have

$$\int \{x \cdot \cos x\}dx = \cos x \cdot \frac{1}{2}x^2 + \int \{\sin x \cdot \frac{1}{2}x^2\}dx$$

which is messier than our starting integral. Now let's follow that sage advice and make

$$u(x) = x, \text{ and } \frac{dv(x)}{dx} = \cos x,$$

which makes $\dfrac{du(x)}{dx} = 1$, and $v(x) = \sin x$.

Plugging these functions into the integration by parts equation, we now have

$$\int \{x \cdot \cos x\} dx = x \cdot \sin x - \int \sin x dx \,,$$

which is readily solvable. Don't forget that when you evaluate your integral over some interval of $x$, you must apply that interval to the $u(x) \cdot v(x)$ part of the integration by parts equation. That is, if you are looking at the interval $a \le x \le b$, then you evaluate $u(x) \cdot v(x)$ at those limits just as you evaluate your newly derived integrals at those limits.

## Example 1: *Root Mean Square*

An electrical engineer wants to calculate the electrical power dissipated[22] in a circuit. Electrical power is the voltage (in volts) applied across the circuit squared, divided by the resistance of the circuit (in ohms). That is,

$$P = \frac{V^2}{R} \,,$$

where

$P$ = power in watts[23]
$V$ = voltage in volts
$R$ = resistance in ohms

Question. If the total resistance in the circuit is 10 ohms and the applied voltage is 155 volts, how much power does the circuit use?

We hope that the first question back to us is: Are you using AC (alternating current) or DC (direct current)?

If we answer DC, then you know that the voltage is constant as though it's being supplied by a battery (at 155 volts, it's an unusual battery, but possible). The answer is readily calculated as:

---

[22] When we say *energy dissipates*, we in no way imply that energy is destroyed. Energy is never created nor is it ever destroyed. All we mean is that the heat and light energy emanating from a resistor, such as a light bulb, is converted to less usable forms of energy. Thermodynamics covers this concept.

[23] Power is the rate of energy use (energy divided by time). A 100-watt light bulb converts 100 joules of energy per second into heat and light.

$$P = \frac{V^2}{R}$$
$$= \frac{155^2}{100}$$
$$= 240 \text{ watts}$$

"That was easy," you say, "just algebra." Yes, just algebra, but it's also electrical engineering.

If, however, we had said AC, then we'd need to know the form of the voltage. Assume that the voltage is represented by the sine function

$v(t) = 155\sin(2\pi \cdot 60)t$ , where $t$ is in seconds.

Now we still have a quandary. The voltage varies from −155 volts to 0 then to +155 volts and back down. And it's making these changes 60 times a second. What do you do?

A purely resistive circuit is not affected by the frequency of the applied voltage, so for this circuit, you don't care about the argument of the sine wave. Also, if you are only looking for the dissipated power, you don't care if the voltage is plus or minus, just as in the DC case you didn't ask which way the battery[24] was connected. All we care about is the effective voltage, but be very careful about that word *effective*. You can look at the sine wave and see that its average value over a cycle is zero, yet you know that the light bulb gets hot and gives off light. What you want is the effective voltage that will dissipate power in the resistor, and you don't care about its sign. You want a value for the equivalent voltage if it were DC. You want the root mean square (abbreviated RMS) for the AC voltage, which is

$$V_{RMS} = \sqrt{\frac{1}{T}\int_0^T v(t)^2 dt} \text{ , where T is the time cycle period of } v(t).$$

Look at the RMS calculation: it makes sense. Squaring the applied voltage eliminates any concern about the sign of the alternating voltage. Integrating over a whole period lets you account for the highs and lows of the varying function, and then you divide by the time period to get the average value. Since that value is a squared term, you must then take the square root to determine how much voltage is heating up your circuit.

[24] Don't show such indifference with your computer. Transistors care very much which way you connect the batteries.

This calculation works for any periodic function (it even works for the DC applied voltage, but that's a lot of work for no gain). You had a sine function applied to our circuit, so the RMS value[25] is

$$V_{RMS} = \sqrt{\frac{1}{T} \int_0^T v(t)^2 dt}$$

$$= \sqrt{60 \cdot \int_0^{1/60} \{155^2 \sin^2(2\pi \cdot 60)t\} dt}$$

$$= 155 \cdot \sqrt{60 \cdot \int_0^{1/60} \{\sin^2(2\pi \cdot 60)t\} dt}$$

If you don't know the integral of the sine squared function off the top of your head, look in an integral table. It is

$$\int \{\sin^2 ax\} dx = \frac{1}{2a}(ax - \sin ax \cos ax).$$

Hence,

$$V_{RMS} = \frac{1}{\sqrt{2}} \cdot 155$$

$$= 110 \text{ volts}$$

which is the effective voltage that is heating your circuit, and the power is

$$P = \frac{V_{RMS}^2}{R}$$

$$= \frac{110^2}{100}$$

$$= 121 \text{ watts}$$

A few things to note about the RMS calculation are:

1) The frequency of the applied function had no effect on its RMS value. A higher frequency just integrates over a smaller interval.

---

[25] A note on units: the frequency of the sine wave is 60 Hz (cycles/second) so, of course, the time period, $T$, is 1/60. Hence, $1/T$ equals 60 with units of 1/seconds. The units of the integral are volts squared (or whatever forcing phenomena squared) times time, the unit that we integrate over. Hence, the time units cancel when we divide by $T$, and we take the square root of the phenomena that we squared, so we are back with the units that we started with.

2) Looking at the symmetry, it should be obvious that we'd get the same RMS value had we used cosine instead of sine to describe our applied function.

3) The result, $V_{RMS} = \dfrac{1}{\sqrt{2}} V_{max}$ only holds for sines and cosines. You must do the calculation if you have a different function.

4) Note that, except in a few cases, the RMS value is not the average value[26] even if the changing voltage never goes negative.

5) In most electrical engineering applications involving power dissipation, AC voltages and currents are stated in their RMS values.

6) The RMS concept is not for electrical engineers only. RMS is used in such diverse applications as mechanical engineers measuring vibration phenomena and acoustics engineers measuring sound pressure.

# Example 2: *Fourier Analysis*

Engineers often work with periodic phenomena that do not follow well-behaved mathematical functions. Some of these phenomena, as modeled, do not even appear to follow our definition of a function[27]. For example, the electronic signal shown in Figure 6-11 is found all over your computers and TV sets. Your boss wants these signals reduced to mathematical functions. How would you mathematically model these signals?

---

[26] As entertainment during your next coffee break, consider a square wave voltage that is equal to zero for half its period and equal to $V_0$ for the other half. If you calculate the power dissipated in the resistor for this varying wave it is zero for the first half of each cycle, and it is $\dfrac{V_0^2}{R}$ for the second half of the cycle. Hence the power dissipated during the period of the square wave is $\dfrac{V_0^2}{2R}$. If you started out the wrong way and observed that the time average voltage is obviously $\dfrac{V_0}{2}$, you might then erroneously say that the dissipated power was $\dfrac{V_0^2}{4R}$. Now, you calculate the RMS value of this square wave and see that the *RMS* isn't *average*.

[27] The plotted phenomena appears to have two different values at one time, which violates the definition for a function. Nothing in our engineering world changes that fast. What we have are changes that take place at tiny time intervals as compared to the time that they are either unchanging or slowly changing within that period. All is relative.

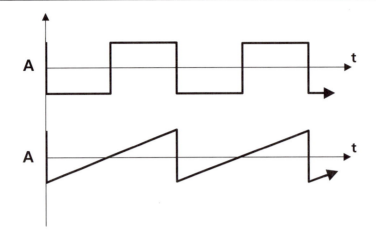

**Figure 6-11. Example of a signal found in computers and TVs.**

Fortunately the French mathematician Jean Baptiste Fourier has shown us a way[28] to model these types of waveforms. His premise is that any periodic phenomena can be modeled by a series of sine and cosine functions. Sounds good, but can we really represent those signals in Figure 6-11 as

$$f(t) = a_1 + a_1 \cos \omega_0 t + a_2 \cos 2\omega_0 t + \cdots + a_n \cos n\omega_0 t + \cdots$$
$$+ b_1 \sin \omega_0 t + b_2 \sin 2\omega_0 t + \cdots + b_n \sin n\omega_0 t + \cdots \quad ?$$

where we use $\omega_0 = 2\pi f_0$ for the argument of the function and the time period of the recurring phenomena is $\frac{1}{f_0}$. For the moment, let's take the liberty of calling that phenomena represented in Figure 6-11 a function, since we're trying to represent it as a series of trig functions. We'll call those waveforms $F(t)$. So let's see if we can represent $F(t)$ as $f(t)$. All we have to do is to determine an endless number of $a_i$'s and $b_i$'s, the coefficients of the terms in Fourier's series.

We don't promise that it'll always be easy, but if we use a property of the integration of sines and cosines over their period, we can get a handle on the problem. Look at a sine or cosine plotted over its period (Figure 6-12), and you see that not only the basic function integrates to zero but that multiples of the function also integrate to zero except when a sine or cosine is multiplied times itself. That is,

---

[28] And though we used an electronics example, Mr. Fourier first published the idea in his treatise, *The Analytical Theory of Heat* (1822).

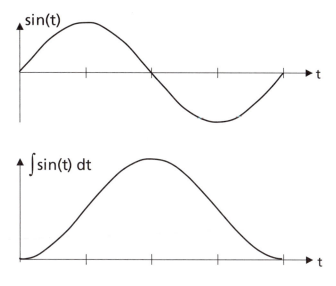

**Figure 6-12. Integrals of sines and cosines.**

$$\int_0^{2\pi} \sin(\theta)d\theta = 0, \quad \int_0^{2\pi} \cos(n\theta)d\theta = 0, \quad \int_0^{2\pi} \{\sin(j\theta)*\cos(n\theta)\}d\theta = 0,$$

$$\int_0^{2\pi} \{\sin(n\theta)*\sin(j\theta)\}d\theta = 0, \quad \int_0^{2\pi} \{\cos(n\theta)*\cos(j\theta)\}d\theta = 0, \text{ where } n \text{ and } j$$

are integers not equal to zero, and $n \neq j$. If, however, $n = j$ then

$$\int_0^{2\pi} \{\cos(n\theta)*\cos(j\theta)\}d\theta = \int_0^{2\pi} \{\cos^2(n\theta)\}d\theta$$

$$= \int_0^{2\pi} \{\sin^2(n\theta)\}d\theta$$

$$= \pi$$

Does this give you a hint on how to proceed? Just follow these steps and we'll see:

1) We'll put $F(t)$, the waveform that we're trying to model, on the left-hand side of the equation.

2) Then we'll put the sine and cosine series on the right-hand side of the equation. That is, we'll assume for the moment that

$$F(t) = f(t).$$

3) Now we'll employ our new trig integration tools. The first thing we'll do is integrate both sides over the fundamental[29] period. As we have seen, all integrals on the right-hand side will be zero except for the $a_0$ term, which will produce

$$\int_0^{2\pi/\omega_0} F(t)dt = 2\pi a_0 .$$

This leaves us having to integrate $F(t)$. We'll look at an easy example after we finish the fundamentals.

4) We'll follow along multiplying both sides by first the fundamental frequency and then by successively higher harmonics. Every time we multiply each side by $\cos(n\omega t)$ and integrate[30] over the fundamental period, we get

$$\int_0^{2\pi/\omega_0} (F(t) \cdot \cos(n\omega t)\}dt = \int_0^{2\pi/\omega_0} a_n \cos^2(n\omega t)dt$$

$$= \frac{\pi}{\omega}a_n .$$

Likewise, if we multiply through by $\sin(n\omega t)$ and then integrate over the period, we get

$$\int_0^{2\pi/\omega_0} (F(t) \cdot \sin(n\omega t)\}dt = \int_0^{2\pi/\omega_0} b_n \sin^2(n\omega t)dt$$

$$= \frac{\pi}{\omega}b_n .$$

And that leaves us having to integrate $F(t)$ to determine $a_0$ and then integrating $F(t)$ with a sine or cosine function for each of the remaining coefficients needed to make up $f(t)$. "Are we better off?" you may inquire.

Well, we are no worse off. How much better we may be depends on the form of $F(t)$. Let's look at the easy example we promised earlier.

---

[29] By fundamental period, we are referring to the period of the lowest (or fundamental) frequency that we are considering—that is, the period of the wave form that we are trying to mathematically model. Integer multiples of the fundamental frequency are called harmonics. The frequency that is twice the fundamental frequency is the second harmonic, etc.

[30] Yes, we changed the appearance of the integrand from $\sin(n\theta)d\theta$ with period 0 to $2\pi$ to an integrand of $\sin(n\omega_0 t)dt$ with period 0 to $2\pi/\omega_0$. Check your trig: it works.

Let's look at the square[31] wave. We'll call the square wave a function and represent it by $F(t)$ which has the value $V_0$ for $0 \leq t \leq \pi/\omega_0$ and the value of zero for $\pi/\omega_0 \leq t \leq 2\pi/\omega_0$, and then it repeats (and repeats). Let's look at the $a_0$ term first (step 3 from above).

$$a_0 = \frac{1}{2\pi} \int_0^{\frac{2\pi}{\omega_0}} F(t)dt$$

$$= \frac{1}{2\pi} \left( \int_0^{\frac{\pi}{\omega_0}} V_0 dt + \int_{\frac{\pi}{\omega_0}}^{\frac{2\pi}{\omega_0}} (0)dt \right)$$

$$= \frac{V_0}{2}$$

for our square wave of amplitude $V_0$.

We'll superimpose our first term for $f(t)$ onto the graph of $F(t)$ in Figure 6-13 where we see that $a_0 = \frac{V_0}{2}$ conveniently cuts $F(t)$ in half. That looks like a good start.

Now, we'll look for the $a_n$ ($n > 0$) coefficients. We know from step 4) that

$$\int_0^{\frac{2\pi}{\omega_0}} (F(t) \cdot \cos(n\omega t)\}dt = \frac{\pi}{\omega} a_n$$

Hence,

$$a_n = \frac{\omega}{\pi} \int_0^{\frac{2\pi}{\omega_0}} (F(t) \cdot \cos(n\omega t)\}dt .$$

Since we know the form of $F(t)$, let's integrate it along with cos(n$\omega$t) to get

---

[31] Of course a square wave is not square. How can it be when the units for values plotted on the x-axis (usually the units are time) have different units than those plotted on the y-axis (which may have units of almost anything)? We call it a square wave because the value is *up* half of the time period and *down* the other half.

$$a_n = \frac{\omega_0}{\pi} \int_0^{\frac{2\pi}{\omega_0}} \cos(n\omega_0)t \cdot F(t)dt$$

$$= \frac{\omega_0}{\pi} \int_0^{\frac{\pi}{\omega_0}} \cos(n\omega_0 t) \cdot V_0 dt + \frac{\omega_0}{\pi} \int_{\frac{\pi}{\omega_0}}^{\frac{2\pi}{\omega_0}} (0) \cdot \cos(n\omega_0 t)dt$$

$$= \frac{1}{n\pi} \Big[ \sin(n\omega_0 t) \Big]_0^{\frac{\pi}{\omega_0}}$$

$$= 0$$

We have just eliminated a bunch of calculations. All of the cosine coefficients are equal to zero when we apply Fourier analysis to a square wave.

Let's look at the sine coefficients. Again from step (4) we have the general form of $b_n$ as

$$b_n = \frac{\omega_0}{\pi} \int_0^{\frac{2\pi}{\omega_0}} (F(t) \cdot \sin(n\omega t))\}dt$$

So let's integrate $F(t)$ along with $\sin(n\omega t)$ to get

$$b_n = \frac{\omega_0}{\pi} \int_0^{\frac{2\pi}{\omega_0}} \sin(n\omega_0)t \cdot F(t)dt$$

$$= \frac{\omega_0}{\pi} \int_0^{\frac{\pi}{\omega_0}} \sin(n\omega_0 t) \cdot V_0 dt + \frac{\omega_0}{\pi} \int_{\frac{\pi}{\omega_0}}^{\frac{2\pi}{\omega_0}} (0) \cdot \sin(n\omega_0 t)dt \ .$$

$$= \frac{1}{n\pi} \Big[ -\cos(n\omega_0 t) \Big]_0^{\frac{\pi}{\omega_0}}$$

Do the algebra/trig and see that

$b_n = 0$ if $n$ is an even number

and

$$b_n = \frac{2}{n\pi} \text{ if } n \text{ is an odd number.}$$

So,

$$F(t) = f(t)$$

$$= \frac{A_0}{2} + \frac{2}{\pi}\left\{ \sin(\omega t) + \frac{\sin(3\omega t)}{3} + \frac{\sin(5\omega t)}{5} + \cdots \right\}.$$

The first few terms are plotted in Figure 6-13. Does it look reasonable?

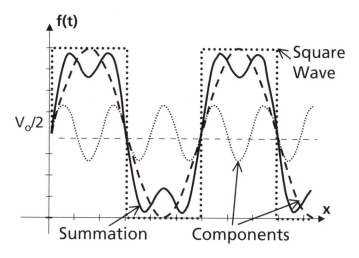

**Figure 6-13. Trig functions and a square wave.**

## Example 3: *Another RMS example*

An ideal diode allows electrons to pass in only one direction. You have a circuit with a diode and a 10-ohm resistor in series with it as in Figure 6-14. Your assignment is to determine the amplitude of the voltage of an applied sine wave if you want the resistor to dissipate 100 watts of power.

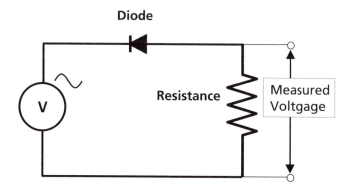

**Figure 6-14. A circuit with a voltage source, a diode, and a resistor.**

Power is the RMS[32] current times the RMS voltage. We, however, only have to perform one RMS calculation because the RMS current is the RMS voltage divided by the circuit resistance. Therefore, as stated

$$P = I_{rms} \cdot V_{rms} \text{, (where } P \text{ is power, } I \text{ is electrical current, and } V \text{ is voltage)}$$

but,

$$I_{rms} = \frac{V_{rms}}{R} \text{ (where } R \text{ is the resistance in ohms)}$$

hence

$$P = \frac{V^2_{rms}}{R},$$

and

$$V^2_{rms} = P \cdot R$$
$$= 100 \cdot 10$$
$$= 1000$$

Hence

$$V_{rms} = \sqrt{1000}$$
$$= 33.3$$

But this is not the RMS value of a sine wave voltage. It's the RMS value of only half a sine wave (don't forget the diode in the circuit) as shown in Figure 6-15. Hence, you must solve for $V(t)$ in the RMS equation:

$$V_{rms} = \sqrt{\frac{1}{T}\int_0^T V^2(t)dt} \text{ (where } T \text{ is the period}[33] \text{ of the sine wave).}$$

---

[32] Normally the voltage and current are stated as the RMS values with no subscripts. We are using the subscripts here since we're using both RMS values and peak amplitudes.

[33] Note that the frequency of the sine wave does not enter the equations other than to determine the period of integration. If we had capacitors or coils in the circuit, the frequency would play a dominant role.

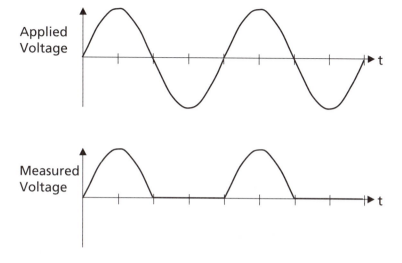

**Figure 6-15. Time dependent voltage of a sine wave
after passing through a diode.**

As we see in Figure 6-15, $V(t)$ is represented by two different functions, depending on what part of the period we are looking at. It is

$$V(t) = V_0 \sin t \text{ for } 0 \le t \le \pi$$

and is equal to zero elsewhere in the period.

Therefore,

$$V_{rms} = V_0 \sqrt{\frac{1}{2\pi} \int_0^\pi \sin^2 t \, dt}$$

$$= V_0 \sqrt{\frac{1}{2\pi} \cdot \frac{1}{2}(t - \sin t \cos t) \Big|_0^\pi} \, .$$

$$= \frac{V_0}{2}$$

Hence the amplitude of the applied sine wave voltage is

$$V_0 = 66.6 \text{ volts}.$$

# Example 4: *Automobile Acceleration*

A popular car is advertised as being able to reach the velocity of 100,000 meters per hour, from a dead stop, in 10.4 seconds. This same car is supposed to be able to brake from that 100,000 meters per hour to a dead stop in 41 meters.

Assume that both the acceleration and the braking are constant (which they aren't), and answer the following questions:

- What is the acceleration needed to gain the 100 km/hour speed?

- How far did the car travel in gaining that speed?

- What is the acceleration upon braking?

- Using your newly calculated value for acceleration, how far would it take you to stop if you were traveling 50 km/hour?

- How long did it take to stop from both of these velocities?

Express the answers in the standard units of meters and seconds.

Let's put the velocities, which are expressed in meters (or kilometers)/per hour, into standard units.

$$100{,}000 \text{ meters/hour} = (100{,}000 \text{ meters/hr})/(3600 \text{ seconds/hr})$$

$$= 27.8 \text{ meters/second}$$

and

$$50 \text{ kilometers/hour} = (50{,}000 \text{ meters/hr})/(3600 \text{ seconds/hr})$$

$$= 13.9 \text{ meters/second}$$

You've seen the pertinent equations earlier in the chapter. They are

$$a = \frac{dv(t)}{dt},$$

$$v(t) = V_0 + \int_0^T a(t)dt,$$

and

$$D(t) = \int_0^T v(t)dt + D_0$$

where

$a$ = acceleration

$v(t)$ = velocity

$V_0$ = initial velocity

$D(t)$ = distance

and

$D_0$ = initial position.

The acceleration of the car is the rate of change of the velocity with respect to time. We can only assume that the acceleration is constant. It probably isn't, but we must work with the data we have, and then compare our calculations with additional data, if we can get it, to refine the values.

With that assumption, we readily calculate the acceleration of the auto as

$$a = \frac{dv}{dt}$$
$$= \frac{27.8}{10.4} .$$
$$= 2.7$$

($V_0$ for this calculation is zero.)

Hence, the acceleration is 2.7 meters/second² or less than one-third the acceleration due to gravity.

During this acceleration period, the car traveled

$$D(T) = \int_0^T v(t)dt + D_0 \text{ meters}$$

We know that $v(t) = at$. We don't care where the car started from, so we let $D_0 = 0$. So, we have

$$D(T) = \int_0^T atdt$$
$$= \frac{1}{2}at^2 \Big|_0^T$$
$$= 146 \text{ meters}$$

To find the acceleration in stopping, let's use

$$v(T) = V_0 + \int_0^T a(t)dt$$

$$= V_0 + aT \quad ,$$

$$= 27.8 + aT$$

$$= 0$$

but we don't know $T$ (the time that it takes to stop).

So let's look at the position integral

$$D(T) = \int_0^T V_0 dt + \iint a(t)dt ,$$

which, assuming a constant acceleration, becomes

$$D(T) = 27.8 * T + \frac{1}{2}aT^2$$

$$= 41 \text{ meters}$$

where the 41 meters was the given value of stopping distance.

Now we have two equations and two unknowns, which we solve to get

$T = 2.9$ seconds.

which leads us to the acceleration of

$$a = -27.8 / 2.9$$

$$= -9.6 \frac{\text{meters}}{s^2} , \text{ or almost the same magnitude as gravity.}$$

Since the acceleration upon stopping is a negative number, we often refer to it as deceleration.

Had the car been only traveling 13.9 m/s (50 km/hr), the time to stop (still assuming constant acceleration of –9.6 m/s²) would have been determined from

$$v(T) = V_0 + \int_0^T a(t)dt$$

$$= V_0 + aT$$

$$= 13.9 - 9.6 \cdot T$$

$$= 0$$

which determines that

$$T = \frac{13.9}{9.6}$$

$$= 1.4 \text{ s}$$

and the distance to stop is

$$D(T) = 13.9 * T - \frac{1}{2} 9.6 \cdot T^2$$

$$= 10.2 \text{ meters.}$$

## Example 5: *Pressure*

Pressure is defined as the force exerted on a surface divided by the area of that surface. A fluid exerts a force on any vessel that contains it. That force is exerted on all of the containing walls. If the vessel is open at the top[34] then the fluid is not externally compressed and the force on the restraining walls is only proportional to the atmospheric pressure and the weight of the fluid. In the examples below we will only consider the pressure from the extra force exerted by the fluid[35] and ignore atmospheric pressure, and we will assume that we are working with a condensed[36] fluid that is not evaporating.

The density of water is 1000 kg/m³, so its weight near the Earth's surface is 9,800 n/m³ (newtons per cubic meter). If we should make a box (we'll ignore its weight) in the shape of a cube with each side one square meter in area and fill it with water, the box would weigh 9,800 newtons[37].

The pressure at the bottom of the box is 9,800 newtons divided by one square meter, or 9,800 n/m². If the box were built to hold the same volume but had a different aspect ratio such that the bottom surface were 0.25 m² in area, the box of water would still weigh 9,800 newtons, but the pressure on the bottom surface would now be 39,200 n/m². The pressure along the sides of both boxes would vary linearly from zero at the top to the maximum (9,800 or 39,200 n/m² depending on which box) pressure at the bottom edge.

The Hoover Dam in Nevada is 221 meters tall. If the lake behind the dam is completely filled, what is the pressure[38] at the bottom of the dam? Of course you can be a wise guy and ask, "Which side of the dam?"

---

[34] The *top* in this discussion is the surface that is furthest away from the center of the Earth.
[35] Engineers call this gauge pressure as opposed to absolute pressure.
[36] By a condensed fluid we mean a liquid and no gases are involved.
[37] One newton is the same as 0.2248 pounds.
[38] We'll ignore the slight compression of the water that'll occur at these pressures.

The answer for the water side of the dam is

P = 9,800 n/m³ · 221 m

= 2,165,800 n/m².

Is the pressure dependent on the surface area of the lake? No, the depth is all that counts. But don't confuse pressure with weight or total force.

Now, let's use calculus to work a pressure problem. Let's help your city design a new aquarium. This aquarium will be a cylinder 40 meters high and 20 meters in diameter. Your job is to determine the total force on the aquarium's surfaces and to determine the force on the rectangular viewing windows.

The force on the bottom of the aquarium is just the weight of the water, which is

$$W = 9800 \cdot \pi R^2 H$$

$$= 9800 \cdot \pi \cdot (\frac{20}{2})^2 40 .$$

$$= 1.2 \times 10^8 \text{ newtons}$$

We'll use cylindrical coordinates[39] to calculate the force on the side walls. As you know, the pressure on the side walls varies from zero at the top to 9,800 × 40 n/m² at the bottom. Therefore, the pressure at any depth (we'll let y = 0 at the upper surface and then let y increase positively as we descend into the tank) is

$P = 9800h$ , where $h$ is the depth into the tank,

and the force on each increment, *dh*, of depth is

$dF = 9800h \cdot 2\pi R dh$ , where $R$ is the radius of the aquarium.

Then the force on the side walls is

$$F = 9800 \cdot 2\pi \cdot 10 \cdot \int_0^{40} h dh$$

$$= 616,000 \cdot \int_0^{40} h dh \qquad .$$

$$= 4.9 \times 10^8 \text{ newtons}$$

The total force on all of the walls is the wall pressure plus the weight on the bottom surface of the tank.

---

[39] We cover cylindrical coordinates in the chapter on vector calculus. Consider this a preview of good things to come.

To determine the force on the rectangular window, we revert to Cartesian nates. The force on a rectangular window is

$$F_{rect} = \int_{h_1}^{h_2} 9800 \cdot w \cdot h \, dh \quad ,$$

$$F_{rect} = \frac{9800}{2} \cdot w \cdot (h_2^2 - h_1^2)$$

where $h_1$ is the depth of the top of the window, $h_2$ is the depth of the bottom of the window, and $w$ is the width of the window.

(We, of course, could have simplified our lives since the pressure is a linear function of depth. We'll let you work out the simpler solutions.)

## Example 6: *Economics*

Your friends in the exotic orchid business want to maximize their profits. They know how to maximize their sales: sell cheap. And they know how to maximize the return on a single sale: ask a high price. But these techniques don't work. If they sell too cheaply, they will not have enough money to pay the mortgage and buy more plants. If they ask too much, they won't sell many flowers it which case they won't have to buy many plants, but they still must pay the mortgage. What to do?

You want to help. You watch their business transactions. You determine that their fixed costs are $1000 regardless of the number of flowers sold, and their per item cost is $50. You also have watched sales and determined that there is a limited demand for these exotic plants and that even if they gave them away, the total that would go out the door in a month is 200 plants, and that sales go linearly to zero as they raise the price to $200 per plant.

Your assignment is to determine the price per plant for your friends to maximize their profits.

Profit, $P$, is equal to income minus costs. The gross income is $x \cdot S$, where $x$ is the number of plants sold and $S$ is the sales price per plant. The costs are the fixed price, $1,000, plus the cost per plant, $50, times the number of plants sold. We'll first determine the number of plants that will be sold when the profit is maximized. Then we'll calculate the asking price.

The profit is the difference between gross income and expenses and its equation is

$$P = x \cdot S - (1000 + 50x),$$

and the price, as a function of number of sales, is

$$S = 200 - 2x.$$

**173**

Combining the two equations we have

$$P = x(200 - 2x) - 1000 - 50x$$

$$= -2x^2 + 150x - 1000.$$

We'll now look for the peaks (where the derivative is equal to zero) by differentiating the profit equation.

$$\frac{dP}{dx} = \frac{d}{dx}\{-2x^2 + 150x - 1000\}$$

$$= -4x + 150.$$

We're in luck. We see that the second derivative of $P$ with respect to $x$ is a negative number. Hence the one peak in this equation is a maximum.

So the profit is maximum when $x = 37.5$. Needless to say, you can't chop these exotic plants in half, so what is it? Is the maximum profit when 37 or 38 plants are sold, and what is the price?

As we said earlier, the price per plant is determined from

$$S = 200 - 2x.$$

So, the number of plants sold, as a function of price, is

$$x = 100 - 0.5S.$$

They'll sell 37 plants if the price is $126, and they'll sell 38 if the price is $124. We put these values of $x$ and $S$ into the profit equation and see that the profit is $1812 with the price of $126, and it's $1862 with the price of $124 per plant.

How else could you advise your friends to increase their profits? That's a question with no easy answers. Can they advertise to increase the demand for their plants, or will the cost of publicity take away any increased income? Can they lower the overhead costs and cost per plant without hurting quality? Sometimes the math is the easiest part of business decisions.

## Example 7: *Semiconductor Photon Detector*

Photons (light waves) are wave-like phenomena and are characterized by the length of these waves. Different colors of light have different wave lengths. The wave length that a photon detector can most easily detect is dependent on the band gap[40] of the semiconductor. The relationship between the semiconductor band gap and the easiest detected wave length is

$$\lambda_p = \frac{1.24}{E_g}$$

---

[40] The band gap is the energy that it takes to break an electron away from binding the atoms together and allowing it to conduct within the detector.

where

$\lambda_p$ is the wave length, in microns (1 micron = $1 \times 10^{-6}$ meters),

and

$E_g$ is the band gap energy in electron volts. One electron volt (eV) is the energy gained by an electron that is accelerated through a potential difference of one volt. 1 eV = $1.602 \times 10^{-19}$ joules.

If you want to detect a certain wavelength and can't find a semiconductor with the requisite band gap, perhaps you can meet your requirements by alloying two different semiconductors. That is, if what you want isn't out there, make your own.

You would start by looking at semiconductors that have lower and higher band gaps and see if the band gap of the mixed crystal (alloy) result is a linear combination of the two. Such materials exist[41] and are used to make detectors.

Remember, you want to detect one $\lambda_p$ which means that you want only one value of $E_g$. Hence you must measure the quantities of your two crystals carefully and mix them completely. How carefully? Your system designer will tell you the allowable error on $\lambda_p$—that is, $(\Delta\lambda)$ that can be tolerated—and you'll use calculus to calculate the allowed compositional error ( $\Delta x$ ).

Look back at

$$\lambda_p = \frac{1.24}{E_g(x)}$$

where $E_g$ is now $E_g(x)$, a function $x$ where $x$ is the proportion of the second crystal material that you added to the first material (if you were adding germanium to silicon you would write the combination as $Si_{1-x}Ge_x$). Now we'll write the equation for $E_g(x)$ as

$$E_g(x) = E_{dif} \cdot x + E_0$$

where $E_{dif}$ is the band gap difference between the two crystals,

$E_0$ is the band gap of the starting crystal,

and $x$ is already defined.

So now we can write the peak detectable wave length as a function of $x$ as

$$\lambda_p = \frac{1.24}{E_g(x)}$$

$$= \frac{1.24}{E_{dif} * x + E_0}$$

---

[41] See "Compositional Control Required in Alloy Semiconductors Used in High Performance Infrared Detector Arrays," A. L. Fripp and R. K. Crouch, *Infrared Physics*, vol. 19, pages 701-702, 1979.

and the change in peak wave length with respect to $x$ is

$$\frac{d\lambda_p}{dx} = \frac{d}{dx}(\frac{1.24}{E_{dif} * x + E})$$

$$= 1.24\frac{E_{dif}}{(E_{dif} * x + E_0)^2}$$

$$= \lambda_p^2\frac{E_{dif}}{1.24}$$

after substituting the equation for $\lambda_p$ into the derivative.

But what we really want to know is how precisely we must control $x$, the composition of our alloy. So let's go from derivatives to differentials and flip the equation over. That gives us

$$\frac{\Delta x}{\Delta\lambda_p} = \frac{1.24}{\lambda_p^2 * E_{dif}}$$

or, to make it more useful

$$\Delta x = \frac{1.24}{\lambda_p^2 * E_{dif}}\Delta\lambda_p.$$

After the system engineers set the desired wave length, $\lambda_p$, and an allowable error, $\Delta\lambda$, you as the semiconductor person can choose the materials that will operate at the stated wavelength. Then you must process the alloying of the materials to maintain the needed $\Delta x$.

A few real-life semiconductors[42] used in such alloying for photon detectors are HgTe alloyed with CdTe and SnTe alloyed with PbTe. The $E_{dif}$ of the first set is 1.87 eV and for the second set it is 0.48 eV. Looking at the equations, we see that the HgTe/CdTe alloy allows for a broader range of detectable wave lengths, but if the SnTe/PbTe alloy meets the wavelength requirement, then the compositional control is less stringent. You're the engineer. You choose.

---

[42] These materials are used to make infrared detectors. Infrared radiation has wavelengths longer than visible light.

# *Probability and Statistics*

*There are three kinds of lies—lies, damn lies, and statistics.* — **Benjamin Disraeli**

The science of statistics, done properly, provides a way to quantify a collection of measurements in an unbiased manner. Often, the "statistics-lies" that we are familiar with are due to the sins of omission rather than the sins of commission. Statistics don't lie, but special interest statisticians can readily mislead the naive. In this chapter we hope to tell the story behind the story so that you'll understand the nature of the questions being asked by the data gatherers and the expected accuracy of such statistics.

Before diving in, we'll first look at the underpinnings of the world of statistics. Probability is the gravity, the force, that holds the parts together. Use probability properly, and statistics is a powerful tool, useful in engineering as well as many other disciplines. As with any powerful tool, use it carefully.

## Probability

Simply defined, probability is the likelihood that something specific will happen. The next level definition for probability is that it's the ratio of the number of ways a certain desired[1] something can happen to the total number of possible outcomes. To expand another small step, look at a single coin toss and assume that you want the coin to land with the head's side showing. As you know, the coin might land with your desired side showing or it might land with the tail's side showing. The probability of getting what you want, the heads showing up, is

$$P_H = \frac{H}{H\_or\_T}$$

$$P_H = \frac{1}{1+1}$$

$$P_H = 0.5 .$$

---

[1] The words *desired outcome* don't have to mean that you want it to happen. We are just referring to the expectation of the occurrence of some given event.

You have one way to get what you want, heads, and the coin has two ways of landing, heads and tails. All this assumes that we have a *fair* toss. That is, we have a coin that is not weighted to land preferentially on only one side, and we have a surface free of gullies or ridges that might force the coin to land on its edge. If that coin is tossed again and again, it will randomly[2] land on heads or tails, and you'll have no advance knowledge of the outcome of the next toss.

Before we go too far, let's clarify some language. When we toss a coin we are conducting an **experiment**. What happens after the toss, that is, whether the coin lands with heads facing up or tails facing up, is the **outcome** of the experiment. We will sometimes refer to the outcome as an **event**.

The other outcome is, of course, that you get a tail, which would have the probable outcome of

$$P_T = 0.5,$$

and using your higher mathematical skills you see that

$$P_H + P_T = 1.$$

Something will happen, and that something will be a head or a tail[3]. Of course, that probability can also be stated as a probability of 100%.

Now might be a good time for you to ask some questions.

*Question #1.* If the probability of a head is 0.5, then why don't we always get a head when we toss twice? After all $0.5 + 0.5 = 1$, doesn't it?

*Answer.* In this game, each toss is a new experiment and past outcomes do not determine future events.

---

[2] As a student of engineering or science, you may not like the idea that you can't control this process. Assuming that you know which side is up when you start the flip, and you know all of the physical parameters of the process (such as coin weight and size, the distance that the coin will drop, the conditions of the surface upon which the coin will land, and weather conditions), you may think you could calculate how much force to put into the initial flip in order to have the coin land on whatever side that you wanted. And you're right. The coin toss is a deterministic process, and all deterministic processes are subject to calculation. The problem is knowing, to the precision needed, all of the initial conditions of the toss. We'll get into such things in the chapter on chaos, but for now, the outcome of the toss is random.

[3] Don't be like our school friend Dufus. Dufus decided to let probability theory determine his study and play habits. Every night he'd toss a coin: heads, he'd go to a movie; tails, he'd go to the tavern. If the coin landed on its edge, he'd study. Maybe Dufus did things right. He has a beautiful tan as he digs those ditches, and we only get to the beach one week a year.

*Question #2.* If we toss a fair coin 100 times, will we get 50 heads and 50 tails?

*Answer.* Maybe, but not likely. You can possibly even get 100 heads or 100 tails, but not very likely (and if you do, we suggest you check the fairness of the coin). You'll probably get something close to 50 heads. Just how close and how much deviation from 50 you can reasonably expect is the purview of statistics, and we'll get there.

*Question #3.* If we toss the coin five times and get five heads in a row, will the *law of probability* dictate that we'll most likely (greater than 50%) get a tail next time?

*Answer.* Whoever coined—pun intended—the phrase *law of probability* should be coerced to read this chapter three times. Sure, you expect a similar number of heads and tails, and five heads in a row is certainly not expected. But an important rule to remember is that in probability, **past events do not impact future events**. You are using a fair coin, and it's been well established that the probability is 0.5 that a single toss will produce a head. Hence the next toss, regardless of past history, is just like the first toss—the probability of a head is 0.5.

*Question #4.* Can the probability of an event be less than zero or greater than one?

*Answer.* No. The probability of a sure thing is one. The coin will land on heads or tails—at least we've never seen a coin land on its edge—so the probability of a head **or** a tail is one. Conversely, if it isn't ever going to happen, the probability is zero, and you can't be more certain of a non-event than that. Be careful using one and zero for describing physical events. Strange things might happen when dealing with the real world. For example: the weather service avoids such statements as "There's a 100% probability of rain today" unless it's raining right now.

*Question #5.* We've heard statements from the weather-predicting folks saying that stronger hurricanes are expected one year compared to another year. Doesn't this violate the "independent event" rule?

*Answer.* Many things that happen do not fall under the guise of a "fair toss." The weather folks look at changing conditions in the atmosphere, water temperatures, and such. Since the conditions for spawning storms change year to year, you might say that the shape of this coin is changing; hence the probability of a given outcome changes. If they make specific predictions far in advance, get a new forecaster.

Let's run through a few more examples and then get formal.

You've tossed a coin once, and you know that the probability of getting a head is 0.5. If you toss it again, you know that the probability of getting a head on the second try is also 0.5. Now we're going to ask: what is the probability of getting three heads when you toss the coin three times? (And please note that tossing a single coin three times is the very same thing as tossing three coins at the same time.)

As we stated earlier, the probability of a desirable event is the ratio of the number of ways that the desirable can occur to the total number of possible outcomes. There is only one way for the desirable event, three heads, to occur. Each toss must produce a head. Now let's count the number of possible outcomes shown in Figure 7-1.

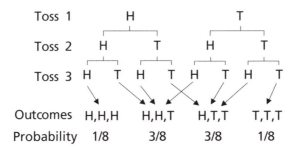

**Figure 7-1. Possible outcomes of tossing a coin twice.**

So we have the one desired outcome divided by eight possible outcomes, hence the probability of getting three heads is 0.125. The exact same argument holds if you want to know the probability of getting three tails, and the answer is the same.

If you have the option, bet on getting a head and two tails. You can toss either H–T–T, T–H–T, or T–T–H to get your desired outcome. You have three ways of getting the head and two tails out of eight possible outcomes. Hence the probability of getting your desired outcome is 0.375. And of course the same argument holds for betting on two heads and one tail.

If we had asked about the chance of getting a tail, a head and then a tail, the probability would have been 0.125 as with the H–H–H possible outcome.

Note all of the possibilities with two tosses:

$P_{H-H} = 0.25$, the probability of getting two heads,

$P_{T-T} = 0.25$, the probability of getting two tails,

$P_{H-T} = 0.25$, the probability of getting a head and then a tail, and

$P_{T-H} = 0.25$, the probability of getting a tail and then a head.

And note that they all sum to unity or 100%, which is to say that one of those outcomes will definitely occur when you toss a coin two times.

From these numbers you can sum the probabilities of the head and a tail

$P_{T-H} = 0.25$ and $P_{H-T} = 0.25$ to get your earlier answer of

$P_{H-T\_or\_T-H} = 0.5$.

*Example*: It is said that Pierre Fermat and Blaise Pascal invented the concept of probability[4] back in 1654. They were playing a dice game and were going for the best three out of five games. Pierre had won two rounds and Blaise had won only one when their game was interrupted. Instead of starting over, they decided to figure out who should have won in the best of five games. With two games left, assuming equal probability of either one winning any specific game, the problem reduces to that of tossing two coins. Pierre would have won the best of five if he won either or both of the subsequent games. Blaise had to win both of the remaining games to get three out of five wins. Hence, there are two ways for Pierre to win and only one way for Blaise to win.

Enough of simple coins, let's roll the dice[5].

Pick up the die and roll it. While it rolls across the floor, you yell, "Three!" So what's the probability of rolling a three when rolling the one die? You have six possible outcomes because the die can stop on any one of its six faces. There is only one way for it to show a three. Hence

$$P_3 = \frac{1}{6}.$$

The same calculation holds for any of the other five possible outcomes.

Now, let's roll the die twice (or roll a pair of dice once—same thing) and look at the possibilities.

You can get a 1, 2, 3, 4, 5, or 6 on the first roll and the same numbers on the second roll. Your possible range of values for the two rolls goes from 2 to 12 with all numbers in between. But the probability of getting any one of those eleven values is no longer the same for each one.

First, determine all the different ways that the dice can provide numbers. The first roll has six possible outcomes, and for each of those six outcomes, the second roll will give six more possible values. Hence, you have 6 × 6 different potential outcomes for your two rolls of the die.

Look at the different ways that the dice can fall.

---

[4] Reference: Science Museum, Ft. Worth Museum of Science and History, April 2002.

[5] As with tossing the coins, the roll of the dice is also a good example of chaos. Rolling of dice is completely deterministic. If you know the exact orientation and momentum of the dice and the exact compliance of the table, then you will know whether you'll lose money in that back alley. However, those variables are not known exactly, and the result of the roll is unknown.

## Table 7-1. Possible sequences of rolling dice.

| First roll | Second roll | Sum | First roll | Second roll | Sum | First roll | Second roll | Sum |
|---|---|---|---|---|---|---|---|---|
| 1 | 1 | 2 | 2 | 1 | 3 | 3 | 1 | 4 |
|   | 2 | 3 |   | 2 | 4 |   | 2 | 5 |
|   | 3 | 4 |   | 3 | 5 |   | 3 | 6 |
|   | 4 | 5 |   | 4 | 6 |   | 4 | 7 |
|   | 5 | 6 |   | 5 | 7 |   | 5 | 8 |
|   | 6 | 7 |   | 6 | 8 |   | 6 | 9 |

| First roll | Second roll | Sum | First roll | Second roll | Sum | First roll | Second roll | Sum |
|---|---|---|---|---|---|---|---|---|
| 4 | 1 | 5 | 5 | 1 | 6 | 6 | 1 | 7 |
|   | 2 | 6 |   | 2 | 7 |   | 2 | 8 |
|   | 3 | 7 |   | 3 | 8 |   | 3 | 9 |
|   | 4 | 8 |   | 4 | 9 |   | 4 | 10 |
|   | 5 | 9 |   | 5 | 10 |   | 5 | 11 |
|   | 6 | 10 |   | 6 | 11 |   | 6 | 12 |

Second, count them: you see the 36 different possible outcomes for the two rolls. Please remember that each of these 36 different combinations have the same likelihood of occurring, but look at the number of different ways to get the different values.

There is only one way for the two rolls of the die to produce a two: each roll must come up as a one. Likewise, there is only one way to produce a twelve: you must roll a six each time. But look at lucky seven. Regardless what number comes up on the first roll, the second roll has a one-in-six chance of summing the two rolls to seven. Hence we have six different ways to roll a seven.

"You speak with forked tongue," our observant reader might remark. "First you say that all rolls have an equal probability of occurring, and then you point out that a player should expect six times as many sevens as twos or twelves."

Not to worry. Just as there is only one way to roll a *one* and then a *one*, there is only one way to roll a *three* and then a *four*, and again, only one way to roll a *four* and then a *three*. The difference between rolling a twelve with the two dice and rolling a seven is that you have a plethora of ways to get that seven.

Let's look at the ways to get each number between two and twelve. (Table 7-2.)

### Table 7-2. Possible values in two rolls of a die.

| Possible values from the two rolls | Number of ways of getting the value |
|:---:|:---:|
| 2 | 1 |
| 3 | 2 |
| 4 | 3 |
| 5 | 4 |
| 6 | 5 |
| 7 | 6 |
| 8 | 5 |
| 9 | 4 |
| 10 | 3 |
| 11 | 2 |
| 12 | 1 |

We generated this table by counting the number of ways of rolling each of the given values as shown in Table 7-1. Note that when you add up all of the number of ways of getting each value, it sums to 36 as should be expected. Nothing else can happen unless a die balances on a point or edge or it splits in two.

The probabilities of rolling each of the values between two and twelve is

$$P_v = \frac{number\_of\_ways\_of\_rolling\_value}{total\_possible\_outcomes}.$$

So

$$P_2 = P_{12}$$
$$= \frac{1}{36}$$

and

$$P_7 = \frac{6}{36} = \frac{1}{6}$$

with the probabilities of the other nine values falling between these two.

This is a good place to introduce the ***frequency function*** and the ***distribution function***. We won't use these concepts for a while, but they fit well with those rolling dice, and we'll have those terms in our pocket for later use. The ***frequency function***, $f(v)$, is the function generated by the probabilities of getting the different possible outcome values, $v$, for whatever operation you're dealing with. In the case of rolling two dice, the frequency function is the probabilities of rolling the different possible values from two to twelve. For example, $f(2) = \frac{1}{36}$ which is the probability of rolling a two. The frequency function for rolling those two dice is shown in Figure 7-2.

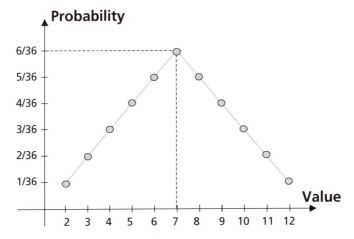

**Figure 7-2. Frequency function for rolling two dice.**

The **distribution function** is essentially the frequency function integrated over the possible outcomes. It's the probability of obtaining a value equal to or less than your desired outcome. The distribution function is defined as

$$F(v) = \sum_{t \leq v} f(t)$$

The distribution function of our rolling dice is plotted in Figure 7-3.

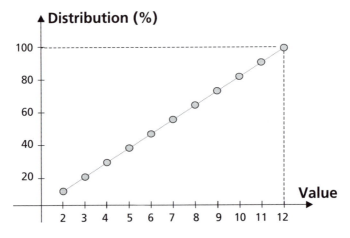

**Figure 7-3. Distribution function for rolling two dice.**

The examples we used in both types of experiments, coins and dice, produced discrete data. If we were not working with something as discrete as values shown on the dice but were measuring the spread of the horns of Texas Longhorn cattle, the accuracy of our data is only limited by our measurement instrument and our courage. We would then have ***continuous*** frequency and distribution functions.

Well, we've tossed coins and rolled dice, and we've seen that determining the probability of a certain event is really a matter of counting. If we are only looking for the likelihood of a certain outcome from the toss of a couple of coins or dice, we can easily count all of the possible ways things can happen and calculate the probability of a certain outcome. At this point you might tell us that the real world isn't so simple and that we can't always count the number of ways something might happen. You might ask, "How would you calculate the probability of an asteroid hitting the earth?"

That's a tough one. Of course we can duck it and say that it's the number of ways of hitting the earth divided by the number of ways of hitting plus the number of ways of missing, but that's a cop out. We won't try to count those ways. Later in this chapter when we get into statistics we'll talk about trying to make do with a sampling of knowledge to determine probable outcomes.

For now, we'll stick with things we can control—at least things we can control on paper. As with the coins and dice, determining probability is largely a matter of counting, so let's re-learn how to count.

We've already seen an example with the dice that the number of ways of rolling two dice is $6 \cdot 6$. Six ways to get a number with the first die, and for each of those six ways there are six ways to get a second number. Should we invent a new game that requires three rolls of a die, then we would have $6 \cdot 6 \cdot 6$ different possible outcomes. Note that we are letting the order of the outcomes of the rolls count. Here, we are not just interested in the final sum produced by the rolls.

We refer to the outcome of rolling

a 2, then a 4, and then a 5

as a different outcome from that rolling

a 4, then a 2, and then a 5

or of rolling

a 1, then a 6, and then a 4

even though all three sets sum up to eleven.

Let's move on to the second lesson in counting. Let's look at the arrangement of things. For example, assume that you and four friends want to use your knowledge of probability. You drive to Las Vegas to roll the dice. (You should expect honest dice there, but you'll still lose.) If each person drives an equal but continuous portion of the trip, how many different ways of taking turns driving might the five of you use on this trip?

There are five different choices in selecting the first driver  After that turn, there are four different ways to choose the second driver, then three ways to get the next driver, then two choices until there is only one person left to finish the trip. So, to put this into an equation,

the number of ways to select drivers = $5 \cdot 4 \cdot 3 \cdot 2 \cdot 1$

= 120.

The math folks call this the **permutation** of a set of objects. The objects here are the five of you driving off to lose your money. The notation and equation for the permutations of $n$ objects taken $n$ at a time is

$$R_n^n = n!$$

where $n!$ is defined as

$$n! = n(n-1)(n-2)\cdots(n-n+1),$$

and for convenience, $0! = 1$, by definition. (We call $n!$ "$n$ factorial.")

Let's take another example. Assume you are in business with five other people. To run the business you need a president, a vice-president, a secretary, and three workers (a stock person, a driver, and a sweeper). Since you are all competent, as well as egalitarian, you decide to rotate the positions. You have six different positions and six people, thus you calculate the permutations of six positions taken six at a time, or

$$R_6^6 = 6!$$

$$R_6^6 = 720,$$

and that's a bunch of ways to run the business.

After following this business model for some time, perhaps you decide that management is not very important, and it doesn't matter who fills the shoes of president and vice-president because they don't do much anyway. You just call those two positions management and let it go as that. Does this decision change the permutation equation? It sure does.

Starting with the workers, you still have 6 people to fill the first worker position, 5 for the second and so on until you get to "management" where the last two people go into those two positions. The permutations are now

$$6 \cdot 5 \cdot 4 \cdot 3 = 360 \,,$$

because there is no difference in how we choose the management positions: we don't have two ways to choose those positions. Or to use fancy notation

$$R^6_{6/2} = \frac{6!}{2!}$$

which is the permutation of six things taken six at a time, but two[6] of the objects are identical.

If we should continue trying to drive our MBA readers to distraction with our innovative business models, then assume that all three worker positions are the same. Our permutations decrease even more.

We will choose the secretary first, and we have six ways of making that choice. Then we are left with five people, two of whom will be managers and three of whom do the production work. We'll now choose a manager from the five remaining people and another manager from the four that are left. The last three people go directly to work and make the company shine.

So now we still have six objects taken six at a time, but two objects are the same, and another three are the same. We have

6 ways to choose the secretary

$$\frac{5 \cdot 4}{2}$$ ways to choose the managers

and only one way to choose the three workers.

So we have

$$\frac{6 \times 5 \times 4}{2} = 60$$

ways to set up your business, which is formally written as

$$R^6_{6/2,3} = \frac{6!}{2!3!}$$

$$R^6_{6/2,3} = 60$$

---

[6]  Yes, we know that 2!=2 and hence the factorial sign is redundant. We write it this way to help us remember to use the factorial operation with larger numbers.

That last example was a bit messy. Let's take a break from business theory and move to pool. If we have six pool balls labeled one through six, how many different ways can we arrange them in a line? We hope that you say the number of permutations is

$$R_6^6 = 6!.$$

Now let's exchange two of the numbered balls for two white cue balls such that we have 6, 5, 4, 3, and two white balls. We can still arrange the balls in $6! = 720$ different ways except that the two white balls are the same, so the net total is $\dfrac{6!}{2}$ or stated as

$$R_{6/2}^6 = \frac{6!}{2!}$$

To continue with the example, let's now reach into other pool tables and wind up with a six, three eight balls, and the two cue balls. That gives us six objects to arrange with one set of two that are alike and another sent of three that are alike. And our permutation equation is

$$R_{6/2,3}^6 = \frac{6!}{2! \cdot 3!}$$

I think that we can see the general equation arising out of the mist. If we have $n$ objects where $r_1$ are of one type, $r_2$ are of another type, ..., and $r_p$ are of the same $p$ type, where

$$r_1 + r_2 + \cdots + r_p = n$$

then the number of indistinguishable arrangements of these $n$ objects is

$$R = \frac{n!}{r_1! r_2! \ldots r_p!} \quad \text{(we dropped the sub- and superscripts to avoid a little mess)}$$

and please note that any, many, or all of these $r_s$ might equal one, which lets this general equation revert to simpler forms.

So far we have used ***all*** of our objects, even if they looked alike, in making the arrangements. Let's go back to our six numbered pool balls and ask how many ways can we put one, and only one, ball in each corner pocket. We are taking six objects and arranging them four different ways. We have six balls to choose among to fill the first pocket, five for the second, four for the third, and lastly, three for the remaining pocket, so

$$R_4^6 = 6 \cdot 5 \cdot 4 \cdot 3$$

or

$$R_4^6 = \frac{6!}{(6-4)!}.$$

And the general equation for arranging $n$ objects $t$ different ways is

$$R_t^n = \frac{n!}{(n-t)!}.$$

If we should only care that four pool balls were in the four pockets and not bother with which one was in the left vs. right corner pocket, we would be talking about the *combinations* of taking those six balls four at a time. It should be obvious that the combinations of choosing the balls is much smaller than the permutations since order does not matter. As before, the permutations for this game is

$$R_4^6 = \frac{6!}{(6-4)!}.$$

But since order doesn't count, the 4! ways of arrangement does not contribute to the combinations of placement. So to get the number of combinations we divide permutations by the factorial of the number of places. That is

$$C_4^6 = \frac{R_4^6}{4!}$$

$$C_4^6 = \frac{6!}{(6-4)!4!}.$$

And the general equation for the combination of $n$ objects taken $t$ at a time is

$$C_t^n = \frac{n!}{(n-t)!t!}.$$

Just for the fun of it, go back to our first pool ball example where you arranged six balls in six different places. How many combinations of these six balls in six places can you get? Of course, common sense tells you that if order doesn't count, you have only one combination of six balls taken six at a time. But this is math, not common sense, so let's do the calculation.

$$C_6^6 = \frac{6!}{(6-6)!6!}$$

$$C_6^6 = 1 \quad (\text{remember, } 0!=1).$$

It looks like common sense might apply to math after all.

Now that you understand permutations and combinations, let's play some more games. We'll put our six pool balls, labeled 1 through 6, into a hat and then pull some out.

*Game #1.* What is the probability of getting a *four* if you reach in and pull out one ball?

You have six different pool balls, taken one at a time, so the number of possible combinations are

$$C_1^6 = \frac{6!}{(6-1)!1!} = 6.$$

You have only one way to get a *six*, so the probability of grabbing that ball is

$$P(4) = \frac{1}{C_1^6}, \text{ or } 1/6.$$

Again, common sense told you that, but it's nice to see the system working.

*Game #2.* If you didn't draw the four ball on the first try, you put whatever ball you drew back into the hat and draw again. What is the probability of getting the four ball this time?

We hope that you immediately say *one-sixth*. If you didn't, then go back and read about *past events not influencing future events.*

*Game #3.* How many times would you have to draw out a ball before you could be sure of getting the four ball?

OK, you're smart and you quickly tell us that we didn't fully state the question. We didn't say if we replaced the balls or not. And that makes a really big difference. So we'll partition Game #3 into Games #3-A, and #3-B.

> *Game #3-A.* We replace the ball each time that we draw out a ball other than the four ball.
>
> Since we've already determined that the probability of getting a *four* (or any other particular ball) is one-sixth on each try, then the probability of ***not*** getting our *four* is five-sixths on each try.
>
> So the probability of not getting the four in two successive tries is
>
> $$P(\#4) = \frac{5}{6} \cdot \frac{5}{6}, \text{ in two tries,}$$
>
> where $P(\#4)$ is the probability of ***not*** rolling a *four.*
>
> This is less than 5/6, but still not good odds.
>
> If we continue like this, we see that in *n* tries the probability of not drawing a *four* is
>
> $$P(\#4) = \left(\frac{5}{6}\right)^n, \text{ in } n \text{ tries.}$$

This probability of the null event (that is, of not getting what you want) gets smaller and smaller as the pre-determined number (*n* in the equation) of tries increases, but it never equals zero. (And please remember, we are talking about the number of tries we plan to make in hopes of drawing the four ball. While in the process of drawing, replacing, and then drawing again, the probability of not getting that four is five-sixths on each turn.)

The probability, of course, of drawing the four ball in n tries is

$$P(4) = 1 - \left(\frac{5}{6}\right)^n.$$

*Game #3-B*. We don't replace the balls as we draw some ball other than a four.

Your chances are much better on this game. Again, your brilliance may beat the drudgery of calculations to the answer, but we'll march through the steps anyway.

This is another example where calculating the null event and then subtracting from one is easier than the direct calculation of the desired event.

Nothing has changed from Game #3-A for the first draw. The probability of not drawing your *four* is still five-sixths. But now we do not replace the drawn ball. On the second turn we are trying to draw the four ball with only five balls in the hat, so the probability of not getting the four ball this time is four-fifths. On the third try the probability of a non-four ball is three-fourths, and then on the fourth it's two-thirds. You see what's happening? Let's write it out.

The probability of not drawing a *four* with five tries is

$$P(\#4) = \frac{5}{6} \cdot \frac{4}{5} \cdot \frac{3}{4} \cdot \frac{2}{3} \cdot \frac{1}{2}$$
$$= \frac{1}{6}$$

Hence, the probability of getting your four ball within five tries is

$$P(4) = \frac{5}{6}.$$

If we went through those five tries and did not drawn a four, we would be left with only one ball in the hat and that must be our elusive four ball. With one ball in the hat, and it must be the four, then the probability of not drawing the four ball this time is zero divided by one which is zero.

Table 7-3 below shows the difference in expected outcomes for the two games.

**Table 7-3. Expected outcomes of picking balls from a hat.**

| | Probability of Drawing a Given Ball if: | |
|---|---|---|
| **Number of Tries** | **Ball Replaced** | **Ball Not Replaced** |
| 1 | 0.17 | 0.17 |
| 2 | 0.31 | 0.33 |
| 3 | 0.42 | 0.50 |
| 4 | 0.52 | 0.67 |
| 5 | 0.60 | 0.83 |
| 6 | 0.67 | 1.00 |

*Game #4.* You have pulled the six ball and then the five ball out of the hat and have not replaced either of them. What is the probability of your continuing this "lucky" streak and pulling, without replacement, the four, three, two, and the one balls out of the hat in that order?

With two balls gone, you now have four different balls in the hat. You have only one way to draw them out in the order that you desire, and you have the permutations of four different things taken four at a time for the possible drawing arrangements of the balls. The probability of drawing your desired set is

$$P = \frac{1}{R_4^4}$$
$$= \frac{1}{4!} \; .$$
$$= \frac{1}{24}$$

If you had drawn the six and then the five ball, don't bet your lunch money that you'll continue drawing the same sequence.

Let's play that game again and approach it in a slightly different way. Assume that the six and five balls are gone. (You know by now that it doesn't matter how they left the hat. All that matters is that they are gone, and we do not have to think about them.) Now let's remove the balls one at time and ask what is the probability of getting our 4, 3, 2, 1 sequence.

The probability of drawing the four on the first try is

$$P(4) = \frac{1}{4},$$

because you have four balls and only one way to draw the four. Likewise, the probability of not drawing the four is

$$P(\#4) = \frac{3}{4}.$$

If you had drawn the four, then the probability of now drawing the three is

$$P(3) = \frac{1}{3},$$

and the probability of not getting the three on this draw is

$$P(\#3) = \frac{2}{3}.$$

You are now down to two balls. This is the last draw. If you draw the two ball, you win, for the outcome of the last draw with one ball is a foregone conclusion. So

$$P(2) = \frac{1}{2},$$

which, of course, is the same as not drawing the two.

Now the probability of getting the 4, 3, 2, 1 sequence is simply

$$P(4,3,2,1) = P(4)P(3)P(2)$$

$$P(4,3,2,1) = \frac{1}{4} \cdot \frac{1}{3} \cdot \frac{1}{2}$$

$$P(4,3,2,1) = \frac{1}{24},$$

which, not surprisingly, is what we calculated before.

"What about the probability of not drawing my desired sequence?" you might ask.

Obviously it's 23/24, but can you calculate it using this stair-step method? Of course you can. When we look at the permutation method, we see 24 possible permutations of which 23 are not your desired sequence. That's the answer. Now let's do it one step at a time.

On the first draw with the four balls, we have already calculated the probability of getting your desired result (drawing the four ball) was 1/4 and the probability of not getting the four was

$$P(\#4) = \frac{3}{4}.$$

If you had drawn the four, then the probability of not getting the three on the next turn was 2/3. So the probability of breaking your sequence on this draw is

$$P(4, \#3) = \frac{1}{4} \cdot \frac{2}{3}$$
$$= \frac{1}{6}.$$

However, even if you had drawn the four ball and then the three ball, you can break your streak on the next draw since you still have two balls in the hat. The probability of either winning or losing at this point is the same. It's 1/2. The probability of getting to this point and then losing is

$$P(4, 3, \#2) = \frac{1}{4} \cdot \frac{1}{3} \cdot \frac{1}{2},$$

$P(4, 3, \#2) = \frac{1}{24}$, (which is the same probability as getting down to two balls and drawing the one you wanted).

The probability of getting your desired sequence is still

$$P(4, 3, 2, 1) = \frac{1}{24}$$

as before, and the probability of not getting there is the sum of the probabilities of getting each of those three undesired paths which is

$$P(\#[4, 3, 2, 1]) = P(\#4) + P(4, \#3) + P(4, 3, \#2)$$

$$P(\#[4, 3, 2, 1]) = \frac{3}{4} + \frac{1}{6} + \frac{1}{24}$$

$$P(\#[4, 3, 2, 1]) = \frac{23}{24}$$

as we already knew.

So in calculating the probability of not drawing the 4, 3, 2, 1 sequence, you have three different routes to get there. Of course, you also had the one path to get your desired sequence. The important point is that each path is valid, and of course, will give the same answer.

*Game #5.* Let's put the pool balls away and go to work. You are now a widget manufacturer, and you are not only concerned about quality control but also about waste of good widgets. You have statistical data showing that 90% of your widgets are good. You don't want to sell defective widgets, so your engineer builds a new testing device that marks bad widgets. After thorough testing, you determine that the new test set is 90% accurate—that is, 10% of the good widgets are marked bad and 10% of the bad widgets are marked good.

*Question #1.* You go to the end of the assembly line and pick up a marked widget. What is the probability that it's really bad? Is your guess 90%? If so, you're wrong.

*Question #2.* Before the test set came into your manufacturing process you shipped 10% bad widgets. What percent of defective products do you now ship? That one is harder to make a bad guess about.

*Question #3.* Are you better off with the new test set in your assembly line?

We built a ***contingency table*** to answer these questions. This is simply a table where we list all possible outcomes of the data and then count the desired events and compare that to all possible events.

*Answer to #1.* It's a 50% chance that the marked widget is really good. Does that seem low? Let's work it out.

Take a typical batch of 100 widgets (which match all of the statistical data). You'd expect to find the following:

**Table 7-4. A contingency table.**

| Good Widgets | | Bad Widgets | |
|---|---|---|---|
| 90 | | 10 | |
| Not Marked | Marked Bad | Not Marked | Marked Bad |
| 81 | 9 | 1 | 9 |

Now you see the 50% probability that a marked widget is really OK. There are 9 + 9 ways of having a marked widget, of which only 9 are bad.

*Answer to #2.* Now that you've made your chart, this one is easy. It's one out of 82 or about 1.2%. You have 81 + 1 unmarked items of which only one is bad.

*Answer to #3.* Yes, you are better off, but are you good enough? We can't answer this one. It depends on the attitude of your customers and what the widget does. You're a smart manager, so perhaps you'll find it worthwhile to send all marked widgets back for more testing to cut down on waste. With your new test set you've gone from shipping 10% defects to just over 1%. If the widget is a vital component of the Space Shuttle, then you must do better.

*Game #6.* Out of approximately 280 million folks in the USA, about 1000 of them are struck by lightning each year[7]. Assuming that these hits are random, what is the probability that you'll be hit in your 80-year life span?

It's easier and more pleasant to determine the probability of not being hit. Since the probability of being hit in any given year is 1000/280 million, then the probability of not being hit is 1–1000/280 million or about 0.999996. The probability of not being hit in 80 years is 0.999996 to the 80th power, or about 0.999714. Now you just subtract the probability of not being hit from one and that is your probability of bad luck during that long life time. It's close to 1 divided by 3500.

# Distributions

We have spent a lot of time on probability games. It's fun, and the thought processes behind these games are germane to understanding statistics and determining when the unscrupulous are misusing this important area of mathematics.

The field of statistics started with the collection of data about things important to the functioning of society. These data counts were taken one by one and then summed into appropriate categories. The Babylonians started collecting data on their crops more than five thousand years ago. They recorded only what they saw and did not try to draw inferences from sampling. Today, we collect what we call vital statistics on such things as births and deaths, which of course, must be counted individually. The United States census is still conducted on a "count only what you see" basis.

There are better ways to use the data. Using probability theory greatly increases the range of statistics. If we understand probability distributions, we can approximate and analyze data, test the reliability of our choice of distribution, and determine the data required for investigating our problem.

Here's a new word: ***distribution***. We all have a basic feel for its broad definition, and in its statistical usage, the word is essentially the same. ***Distribution*** is just how you can expect the outcomes (results) of given events to occur. You've already seen examples of distributions with the coin tosses and the rolling of the dice. These are examples of the ***binomial distribution***.

A binomial distribution is one in which something either happens or it doesn't happen: you toss two heads or you don't toss two heads, or you roll a seven with two dice or you don't roll a seven. We'll soon bore you with details about the binomial distribution, but first let us mention another type of distribution about which we'll

---

[7] That's about the same number that die by accidental electrocution in their homes.

also wax pedantic: the normal, Gaussian, or bell-shaped distribution. For this last one, we'll honor Carl Fredrick Gauss, the great nineteenth-century mathematician and physicist, who first came up with this distribution. The Gaussian distribution complements the binomial distribution as the probabilities of occurrence are not far apart, but it also stands alone in describing physical phenomena. We'll start with the binomial and then introduce other distributions.

## The Binomial Distribution

As we just mentioned, the results of rolling a die follow a binomial distribution. Let's extrapolate on what we've learned to develop some general equations. If we roll the die six times and ask the probability of rolling a three on two, and only two rolls, we know that one sequence for this desired result is 3-1-3-6-4-5. But we don't care what numbers are rolled if they're not 3s. We can just as well write the sequence as 3 - 3 #3 #3 #3. The probability of rolling this sequence is

$$P(\text{3-3 #3 #3 #3 #3}) = \frac{1}{6} \cdot \frac{1}{6} \cdot \frac{5}{6} \cdot \frac{5}{6} \cdot \frac{5}{6} \cdot \frac{5}{6}$$

or to use the general terms

$$P(of\_x\_positive\_events) = p^x q^{n-x}$$

where

$p$ = the probability of getting a desired result ( a three in this example)

$q$ = probability of not getting your desired result (any number except three)

$x$ = number of times you want that certain result (two in this example)

$n$ = number of tries (six in this example)

and don't forget that

$$p + q = 1.$$

Don't despair when you determine that the probability of getting that particular sequence is just a bit more than one in a hundred because there are other sequences that'll give you those threes that you want (we don't care where they occur in the sequence). How many sequences can we count? That's a question for the permutations equation. We want the number of permutations of rolling two threes and four of anything else. That is, we want the number of permutations of six things taken six at a time with two of the six alike (the threes) and the four other things also alike (that is, anything but three). So

$$R_{2,4}^6 = \frac{6!}{2! \cdot 4!}$$

From which we can write the general equation for the permutations as

$$R^n_{x,(n-x)} = \frac{n!}{x!(n-x)!}.$$

And now we see that we have fifteen different ways to roll two things alike. Calculate the factorials if you don't believe us.

The general equation for a given outcome of $x$, and only $x$, desired results in $n$ tries is

$$f(x) = \frac{n!}{x!(n-x)!} p^x q^{n-x},$$

where $f(x)$ is the frequency function for the general form of the binomial distribution.

In the seventeenth century Jacob Bernoulli, from Switzerland, developed the theory for the binomial distribution.[8] Why he isn't given credit for his contribution, we don't know—maybe his name was too hard to spell. Where does the name "binomial" originate? That one we can answer. Using techniques developed for polynomials makes working with the "Bernoulli" distribution much easier.

Back to math. If you take a binomial expression, such as $(p+q)$ and raise it to the power $n$ (where $n$ is a positive integer or zero) you get

$$(p+q)^n = p^n + np^{n-1}q + \frac{n(n-1)}{2}p^{n-2}q^2 + \ldots + npq^{n-1} + q^n.$$

If you look carefully, you'll see that we can write this expansion as

$$(p+q)^n = \sum_{x=0}^{n} \frac{n!}{x!(n-x)!} p^{n-x}q^x$$

which is, term by term, the same thing as the frequency function for our binomial/Bernoulli distribution, or

$$(p+q)^n = \sum_{x=0}^{n} f(x).$$

---

[8] 'Twas a smart family. His brother Johann performed pioneering work in the development of calculus, and his nephew Daniel anticipated the law of conservation of energy.

By using the binomial expansion, we have the terms of $f(x)$ for all possible values of $x$. In our earlier example where we were seeking the probability of rolling three twice, and only twice, in six rolls of the die, we can now look at the binomial expansion of

$$\left(\frac{1}{6} + \frac{5}{6}\right)^6$$

to determine the probability of rolling a three 0, 1, 2, 3, 4, 5, or 6 times.

Side note: Look at

$$R^n_{x,(n-x)} = \frac{n!}{x!(n-x)!} \quad \text{and} \quad R^{n+1}_{x,(n+1-x)} = \frac{(n+1)!}{x!(n+1-x)!}$$

and note that

$$R^{n+1}_{x,(n+1-x)} = \frac{(n+1)}{(n+1-x)} * R^n_{x,(n-x)}.$$

"Cute," you say. "You have one value calculated and *Bingo!* with a simple manipulation you have the next one, and on you go without a lot a messy factorials. But why bother? We have computers to handle this stuff."

Sure we have computers and we'd never ask you to work more than one or two factorials. But what we have here is a ***recursion formula***. This particular one allows us to find all of the coefficients of a binomial expansion (and we have an even easier way which we'll show you at the coffee break). Recursion formulas find many usages in computer programming. Now you've at least heard the term.

# Pascal's Triangle

Blaise Pascal, a seventeenth-century Frenchman, not only helped invent probability but also made contributions to fluid mechanics. Some say he invented the concept of the computer. For us, he invented a useful triangle for determining the coefficients of the binomial expansion.

Note that

$(p+q)^0 = 1$         (the coefficient is 1)

$(p+q)^1 = p + q$      (the coefficients are still 1)

$(p+q)^2 = p^2 + 2pq + q^2$    (coefficients are 1, 2, 1)

and

$$(p+q)^3 = p^3 + 3p^2q + 3pq^2 + q^3 \quad \text{(coefficients are 1,3, 3, 1)}$$

and so on.

Pascal looked at this recurring pattern and came up with his famous triangle.

**Table 7-5. Pascal's triangle.**

|   |   |   |   |   |   |   |   |
|---|---|---|---|---|---|---|---|
|   |   |   | 1 |   |   |   | n = 0 |
|   |   | 1 |   | 1 |   |   | n = 1 |
|   | 1 |   | 2 |   | 1 |   | n = 2 |
| 1 |   | 3 |   | 3 |   | 1 | n = 3 |
| 1 | 4 |   | 6 |   | 4 | 1 | n = 4 |
| 1 | 5 | 10 |   | 10 | 5 | 1 | n = 5 |
| 1 | 6 | 15 | 20 | 15 | 6 | 1 | n = 6 |

where we see the numbers in row by succeeding row are generated by adding the numbers to the immediate left and right above them (see Table 7-5). For example, the third term from the left in the row for n = 6 is determined by adding the 5, above left, to the 10, above right. If there's an elegant theory for this triangle, we don't know what it is, but it works.

Isn't Pascal's triangle neat? You not only have a method to easily determine the coefficients for binomial expansions, but you also have the permutations for *n* items taken *n* at a time where *x* and (*n* – *x*) of the items are alike. Do you remember that we calculated 15 different ways you could have two, and only two, *threes* in the arrangement of six dice? Well, there it is. Go down[9] to the row for *n* = 6, and you see that the triangle tells you there is one way to have no *threes*, six ways to have one *three*, fifteen ways to have two, and so on. Of course, the magic number doesn't have to be three. The same procedure is used, and the same values are obtained, if we look for any specific number on the die.

The same theory also works if we're talking about tossing coins and looking at the number of ways of getting a given number of heads. If you toss your coin six times, you have one way to toss zero heads, six ways to toss one head, fifteen ways to toss two heads, and so on.

---

[9] Sure, we start with *n* = 0, which makes the triangle work. It makes some sense in the binomial games, $(p+q)^0 = 1$, so don't worry about taking zero things zero at a time.

Please note that the triangle only gives you the **number of ways** of obtaining your desired result. It does not give you the **probability** of obtaining that result. Pascal gives you the coefficients to the $p^{n-x}q^x$ term. For example, the probability of rolling two, and only two, *threes* in six turns is 0.201, and the probability of tossing two and only two heads in six tosses of the coin is 0.234 (both to three decimal places). Those two probabilities seem close when the probability of rolling a three is so much smaller than tossing a head. Look, though, at the probabilities for all possible occurrences of the rolls and tosses. Table 7-6 and Table 7-7 list your chances of getting 0, 1, 2, 3, 4, 5, or 6 of any given number on the die or any chosen side of the coin.

### Table 7-6. Die roll probabilities.

| x | $R^6_{x,6-x}$ | Probability where p=1/6, q=5/6 |
|---|---|---|
| 0 | 1 | 0.335 |
| 1 | 6 | 0.402 |
| 2 | 15 | 0.201 |
| 3 | 20 | 0.054 |
| 4 | 15 | 0.0080 |
| 5 | 6 | 0.0006 |
| 6 | 1 | 2.14E-05 |

### Table 7-7. Coin toss probabilities.

| x | $R^6_{x,6-x}$ | Probability where p=q=1/2 |
|---|---|---|
| 0 | 1 | 0.016 |
| 1 | 6 | 0.094 |
| 2 | 15 | 0.234 |
| 3 | 20 | 0.313 |
| 4 | 15 | 0.234 |
| 5 | 6 | 0.094 |
| 6 | 1 | 0.016 |

Figure 7-4 shows the differences in the possible outcomes quite clearly. The dice curve is quite asymmetrical while the coin curve is symmetrical and almost resembles a sun hat or maybe a bell in its shape. Remember this bell-shaped curve; we'll make a lot more of it later.

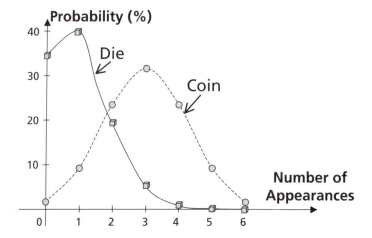

**Figure 7-4. Probabilities of rolling a chosen number *x* times in 6 tries when rolling dice and of tossing *x* heads in 6 tosses of a coin.**

You've seen that the probability of rolling the number 3 six times in a row is a very small number, something like twenty-one parts in a million. And, of course, the same holds for any number on the die just as long as that number is chosen before the first roll. Now, let's have a quiz. You have made one roll of the die. What, now, is the probability of rolling that same number six times in a row? The answer is below[10].

# Moments

Now we look at moments. The statisticians are stealing this term from the physicist, and the physicist uses it to describe the distribution of mass around some point of reference. Though there are higher orders that some folks want you to suffer through, we'll only talk about first and second moments.

---

[10] The answer is $(1/6)^5$, approximately one out of ten thousand, the same as rolling a die five times and getting the same predetermined number on all five rolls. In this case, the specific number was determined by the first roll.

We're going to diverge off into physics for a bit. The intent is to use the analogy between statistics and physics. Wade though this and physics will be easier. Skip this introduction for now, go study physics, and then this section will be a piece of cake. (In college I always felt that I needed to have taken all of the other courses before I took any of the courses. I guess that's the beauty of education—everything builds on everything else.)

If you don't want to work on physics right now, skip on down to **Moments in Statistics**.

To the physicist, the first moment about some reference point is the sum of the distances from that reference point times the mass found at that distance. That is

$$\mu_{ref} = \sum_{i=1}^{n} x_i m_i(x_i)$$

where:

$x_i$ = distance from the chosen origin to the $m_i$ mass particle

$m_i(x_i)$ = mass found at position $x_i$.

and

$n$ = number of mass particles

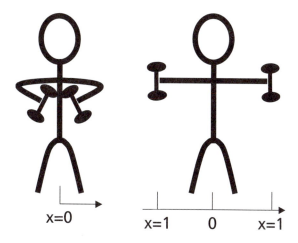

x=0          x=1     0     x=1

**Figure 7-5. First and second moments.**

Our physicist can then divide $\mu_{ref}$ by the total mass and get the *center of mass* of his objects. That is

$$\mu = \frac{\sum_{i=1}^{n} x_i m_i(x_i)}{\sum_{i=1}^{n} m_i(x_i)}$$

Note that if the physicist chose the center of mass, $\mu$, as the origin then

$$\mu_{cm} = \sum_{i=1}^{n}(x_i - \mu)m_i(x_i) = 0$$

where

$\mu_{cm}$ = the first moment about the center of mass

and

$\mu$ = the position of the center of mass.

For every distance and mass seen in one direction there will be an equal set in the opposite direction to cancel it out. That's the definition of center of mass.

To add to the confusion, please note that $\mu_{ref}$ and $\mu_{cm}$ have units of length multiplied by mass while $\mu$, the center of mass, has units of length only.

The center of mass concept is important to the physicist, and it's also important to statisticians. Be patient, you'll see.

In the opening paragraph of this section, we promised you first and second moments. The number comes from the power of the distance in the summation. It was to the degree one, hence first moment. From the first moment we got the center of mass. Think about that first moment for a bit. If we halved each mass element and doubled each distance from the original center of mass of those mass elements, we would not change the value of either the first moment about the center of mass or the position of the center of mass. Yet you know that handling this new distribution will be quite different from the old one and that's where the ***second moment*** comes to the rescue.

The second moment is defined by the exponent 2 on the length term in the equation

$$\mu_2 = \sum_{i=1}^{n} x_i^{2} m_i(x_i).$$

Now you can see that the distance or spread, of the various mass particles, from the reference point has a greater effect on the value of the second moment than the mass itself. The second moment, often called the moment of inertia, is very important to our physicist friend working with any kind of rotational dynamics. (See Figure 7-5.)

## Moments in Statistics

The concept of the first and second moment allows the physicist to make useful characterizations of a mass distribution, and we have been looking at the probability distributions of rolling a given number and of tossing whatever number of heads. Let's see what we get when we apply moments to our probability distributions.

For starters, we'll look at the first moment of the coin toss. That's where we toss a coin six times and determine the probability of getting 0, 1, 2, 3, 4, 5, or 6 heads.

The probability of rolling $x$, and only $x$, heads in six tosses is

$$f(x) = \frac{6!}{x!(6-x)!} \cdot \left(\frac{1}{2}\right)^x \left(\frac{1}{2}\right)^{6-x}$$

where $x$ is the number of positive events we seek and which range from one to six.

The first moment of this distribution[11] is

$$\mu_1 = \sum_{x=0}^{n} x f(x).$$

Looking back at Table 7-1 and multiplying each probability term by its appropriate value of $x$ we have

**Table 7-8. First moment of tossing coins.**

| x | f(x) | x*f(x) |
|---|---|---|
| 0 | 0.016 | 0 |
| 1 | 0.094 | 0.094 |
| 2 | 0.234 | 0.469 |
| 3 | 0.313 | 0.938 |
| 4 | 0.234 | 0.938 |
| 5 | 0.094 | 0.469 |
| 6 | 0.016 | 0.094 |
| Sum | 1 | 3 |

where the first moment, $\mu_1$, is $\mu_1 = 3$.

---

[11] We can just sum over $x$ instead of $x_i$ because there is a $f(x)$ associated with each value of $x$. Also note that when we summed over $i$, we started the summation at $i = 1$ because we started with the first mass particle. When we sum over $x$ the range of possible values, we start the summation at zero because zero is a possible value of $x$.

If we take the next step and try for the center of probability, we just divide $\mu_1$ by the sum of the $f(x)$'s, which of course is equal to one, since the sum of the probabilities of all possible outcomes is one. So the first moment, about zero, of this distribution and the effective center of mass is the same thing. In statistics we call $\mu_1$ (Greek mu), the arithmetic mean, or just the mean[12], of the distribution of possible outcomes.

If we calculate the expected distribution of the number of heads obtained by tossing the coin fifty times, $\mu_1$ would equal 25.

Please note that you should not expect to toss that many heads when playing the game. Do the math. The probability of tossing three heads out of six is

$$f(3) = 0.313,$$

and if we go to 50 tosses

$$f(25) = 0.112.$$

Hence, the probability of tossing the average value, $\mu_1$, decreases as the number of tosses increases. From here on, we'll refer to $\mu_1$ as plain old $\mu$.

Just for the fun of it, decide that you want an 80% probability of tossing the number of heads within some range when tossing six coins. Look at Table 7-7. Sum the probabilities for tossing 2, 3, or 4 heads. The probability of getting within this range is 78%, which is as close to 80% as the distribution of six can get. When you toss the fifty coins you must include 21 through 29 to reach 80%. Do you suppose that the second moment can tell us something about the *spread* of the probabilities about the mean?

The second moment[13] about the mean is

$$\mu_2 = \sum_{x=0}^{n} (x-\mu)^2 f(x).$$

When we run through the numbers we see that $\mu_2$ equals 1.5 for the six coins and 12.5 for the fifty tosses.

---

[12] Other terms of this sort that you ought to know are *mode*, *median*, and *geometric mean*. The mode is the value of which the highest number of events occur. In rolling two dice, the mode is seven since it will occur the most number of times. The median is the center of the frequency distribution, just as in real life—it's the center of the road. In rolling that pair of dice, 7 is the median since there are five possible numbers on either side of it. The median does not have to be a possible value within the distribution.

The geometric mean is a bit trickier. It's defined as $\sqrt[n]{y_1 y_2 ... y_n}$. "What's it good for?" you ask. One use is in determining the effective average interest rate in a succession of compounding payment periods with varying rates, where $x_i$ is 1 plus the interest rate for the $i^{th}$ period. We'll leave it at that.

[13] You will sometimes see the second moment referred to as the *variance* of the data.

We guess that you've already made the connection that $\mu_2 = \frac{1}{2}\mu$ but don't get carried away with that observation. It only holds true for a binomial distribution where the probabilities are equal. Go through the examples that we'll discuss, and you'll see that $\mu_2 = npq$ for a general binomial distribution.

We're sure that you don't see any wonderful simplifying equations coming out of $\mu$ and $\mu_2$, and at this stage, please believe that we didn't either. But Mr. Gauss applied his genius and came up with

$$f(x) = \frac{1}{\sigma\sqrt{2\pi}} e^{-\frac{1}{2}\left(\frac{x-\mu}{\sigma}\right)^2}$$

where $\sigma = \sqrt{\mu_2}$ .

This is his very useful Gaussian distribution[14].

Let's plot $f(x)$ vs. $x$. That is the probability of tossing $x$ heads in $n$ tosses, as calculated from

$$f(x) = \frac{n!}{x!(n-x)!} \cdot p^x q^{n-x}, \text{ where } p = q = \frac{1}{2}$$

and from

$$f(x) = \frac{1}{\sigma\sqrt{2\pi}} e^{-\frac{1}{2}\left(\frac{x-\mu}{\sigma}\right)^2} .$$

Figure 7-6 shows the comparison for $n = 6$. "Not a terrible fit," you might say, and Figure 7-7 is the same comparison for $n = 50$, to which you might comment, "But I see only one curve." That's right. The data points are that close.

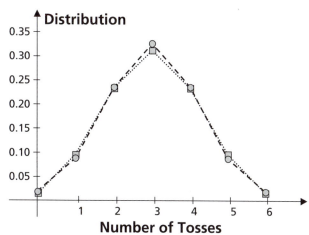

**Figure 7-6. Binomial and Gaussian distributions for six coin tosses.**

---

[14] The Gaussian distribution is also called the normal or bell-shaped distribution.

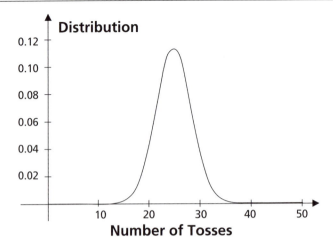

**Figure 7-7. Binomial and Gaussian distributions for 50 coin tosses.**

Your next question might well be, "Does the Gaussian curve fit all binomial distributions?" The answer is, "No." We picked the best fit, $p = q = \dfrac{1}{2}$, for this

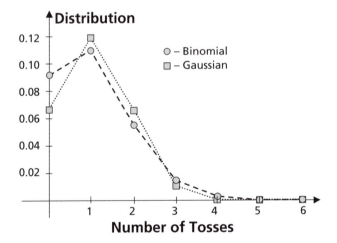

**Figure 7-8. Binomial and Gaussian distributions for six dice rolls.**

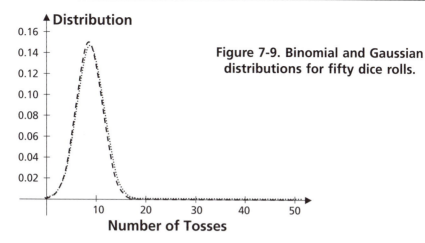

**Figure 7-9. Binomial and Gaussian distributions for fifty dice rolls.**

demonstration. Figure 7-8 and Figure 7-9 plot the direct calculations of $f(x)$ vs. Gauss's method for rolling the dice, and it's still a reasonably good fit.

If, however, we invent another game with a low probability of success, will Gauss still work for us? Let's cut our die into some weird pattern with 25 faces on it and paint each face pink except for one green face. We now have a 25 to 1 chance of rolling a green. We'll roll six times and calculate the probability of the number of greens using the binomial equation and the Gaussian distribution. Figure 7-10 and Figure 7-11 show that the Gaussian does not work as well as it did with the more even probabilities. What now? Luckily the French mathematician Siméon-Denis Poisson[15] has already come to our rescue with his equation. His equation is helpful when we have distributions with large numbers of samples but low probabilities for a successful outcome (that is, a green face with our new rolling game).

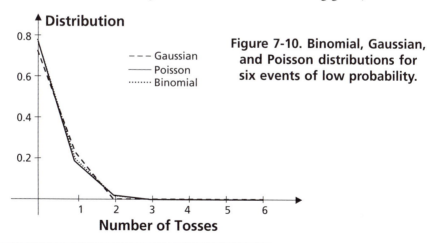

**Figure 7-10. Binomial, Gaussian, and Poisson distributions for six events of low probability.**

---

[15] I'm sure that all of the French students have translated *Poisson* to *fish* in English. The name is real and is very appropriate. We go fishing a large number of times and the probability of catching dinner is very small. Our fishing data fits his distribution.

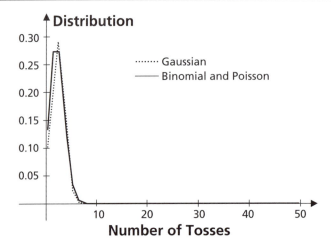

**Figure 7-11. Binomial, Gaussian, and Poisson distributions for fifty events of low probability.**

The *Poisson distribution* equation is

$$f(x) = \frac{e^{-\mu}\mu^x}{x!}.$$

We also plotted the Poisson distribution for these numbers ($n = 6$ and $50$) and probabilities ($p=1/25$ and $q=24/25$) in Figure 7-10 and Figure 7-11. We see that when we have a low probability of success, even with low $n$, Poisson's equation gives a better fit than Gauss's. The difference in fit becomes even more pronounced at higher $n$.

## Probability Distributions in Real Life

Why do we care about these theoretical distributions for binomial probabilities? If we know $p$ and $q$, the probabilities of event occurrence, why don't we just dump it into our little computer and crank away?

The binomial was just an easy way to introduce the Gaussian and Poisson distributions, and these distributions, especially Gauss's, are often quite useful in describing the real world whether we're talking about the spread of a Texas longhorn's horns, the rainfall over a certain area, or a grade distribution in college. In the real world, you'll gather data and then try to fit it to a distribution function. And in that world you're not stuck with just these two. There are more statistical distributions out there than Good-Time U has sororities and fraternities, but you'll have to turn to a more specialized book to learn about them. We haven't finished with the Gaussian, however.

If we look at the equation

$$f(x) = \frac{1}{\sigma\sqrt{2\pi}} e^{-\frac{1}{2}\left(\frac{x-\mu}{\sigma}\right)^2}$$

we easily see that the exponential part gives us the shape of the curve. When applied to the binomial distribution, the pre-exponential part is just a normalizing factor to make the sum over all $n$ equal to unity. When considering all $n$, the probability of something happening is just that: one.

If we're doing something really interesting such as measuring those longhorn's horns, we would likely have the width of the horns on the $x$-axis and number of cattle on the $y$-axis. With that distribution, we'll need a different pre-exponential factor. Also if you're talking about measurements such as this, you're no longer talking about discrete data, such as with tossing coins. You are now handling continuous data. You've gone from summations to calculus.

> This is probably a good time to throw in a short talk on measurements. When we're measuring those horns, how accurately do we measure them? With most of those guys and gals (and incidentally, the gal's horns are generally longer than the guy's), I'd prefer to measure the width using calibrated binoculars, but that won't do. If the cattle would hold still and we used a laser interferometer to its full capability, we might see the horns grow while making the measurement. We'd have an accuracy that isn't needed, and the data would be awful to work with.
>
> So what do we do? We use intervals. And this takes some judgment. Look at the data, determine the range of measurements, and then divide that range into equally spaced intervals and plot the number of longhorns that fall into each interval. Your need for accuracy has been reduced (and your safety increased) and your data handling simplified.
>
> Have we just gone back to discrete data because of the finite number of intervals? No, the data is still continuous as you likely have measurements all across the interval. You plot the function through the center of each interval.

Whether tossing coins or measuring horns, we see that the shape of the Gaussian distribution is characterized by the values of two numbers for that distribution: $\mu$, which is the mean, and $\sigma$ (Greek sigma), which we haven't given a name yet. Let's go with convention and call $\sigma$ the ***standard deviation***. Now let's see why it deserves special recognition.

Looking at the bell-shaped Gaussian distribution, we see that it has one maxima. On either side of that maxima the values trail off but never reach zero. We also see that the curve has two inflection points where, starting from the left, the slope of the curve stops increasing and then starts decreasing before it reaches the maxima. Since the curve is symmetric, we get another inflection point on the right of the maxima. Let's calculate these points and see if they're interesting.

We told you that calculus is useful, and we'll use it here.

To find the maxima, we take the first derivative of Gauss's function

$$\frac{d}{dx}f(x) = \frac{d}{dx}\frac{1}{\sigma\sqrt{2\pi}}e^{-\frac{1}{2}\left(\frac{x-\mu}{\sigma}\right)^2}$$

$$= \frac{-(x-\mu)}{\sigma^2}f(x)$$

and determine where it equals zero. We find that we have only one maxima which occurs at $x = \mu$, the average value of the distribution. That's another way of saying that most things tend to the average value. How close to the average value will depend on the position of the inflection points, which we'll find by taking the derivative of the first derivative and determining its zeros.

$$\frac{d^2}{dx^2}f(x) = \frac{-1}{\sigma^2}\frac{d}{dx}\{(x-\mu)f(x)\}$$

$$= \frac{+1}{\sigma^2}\{1 - [\frac{(x-\mu)}{\sigma}]^2\}f(x)\}$$

And we find our inflection points at

$$x = \mu + \sigma$$

and

$$x = \mu - \sigma.$$

We've been giving hints that the second moment, or as we're using it, the square root of the second moment, relates to the spread of the function. Let's quantify that spread.

Earlier we counted the number of events needed to have an 80% probability of success. Since that was with a discrete function, counting was OK although it could get very laborious if we are dealing with large numbers. Instead of looking at the 80% probability limits, we are going to determine the probability of finding any given value between $x = \mu + \sigma$ and $x = \mu - \sigma$. That is, we are looking at the percentage of the distribution that lies within $\sigma$ of the mean value, $\mu$. Since we are now dealing with a continuous function, we must count by using integral calculus.

The probability that $x$ will fall between $\mu - \sigma$ and $\mu + \sigma$ is

$$\int_{\mu-\sigma}^{\mu+\sigma} f(x)dx = \frac{1}{\sigma\sqrt{2\pi}}\int_{\mu-\sigma}^{\mu+\sigma} e^{-\frac{(x-\mu)^2}{\sigma}}\,dx \ .$$

If you can't find the solution for this integral in your tables, you can do the following:

let $t = \dfrac{(x-\mu)}{\sigma}$,

then $dx = \sigma dt$

and the range of integration becomes $-1 \le t \le 1$.

You should find the new function

$$\frac{1}{\sqrt{2\pi}} \int_{-1}^{1} e^{-\frac{t^2}{2}} \, dt$$

in your integral tables. If you still can't find the solution in the tables, then flip the function about $t = 0$ (the function is symmetric about its center) and look for

$$\frac{2}{\sqrt{2\pi}} \int_{0}^{1} e^{-\frac{t^2}{2}} \, dt .$$

Or you can take our word for it that

$$\int_{\mu-\sigma}^{\mu+\sigma} f(x)dx = \frac{1}{\sigma\sqrt{2\pi}} \int_{\mu-\sigma}^{\mu+\sigma} e^{-\frac{(x-\mu)^2}{\sigma}} \, dx = 0.68 ,$$

to two decimal points.

If you're lazy and smart, you can also take our word that

$$\int_{\mu-2\sigma}^{\mu+2\sigma} f(x)dx = \frac{1}{\sigma\sqrt{2\pi}} \int_{\mu-2\sigma}^{\mu+2\sigma} e^{-\frac{(x-\mu)^2}{\sigma}} \, dx$$

$$= 0.95$$

also to two decimal points.

Hence, we see that approximately two-thirds of the population of whatever we're looking at falls within +/– 1 $\sigma$ and nearly 100% falls within $2\sigma$ of the mean value.

Using $\sigma$ as a measurement of spread is so convenient that it's often called the ***unit of standard deviation***.

# Correlation

When looking at pairs of data such as the amount of rainfall and the number of orchids grown in your garden plot, you would expect a relationship between the two. But watch the language. In statistics we don't use touchy-feely words like relationship. We ask for the ***correlation*** between the two variables, and since we're talking math, we quantify the correlation between the two.

Let's assume that you've kept the following monthly data over the past two and a half years. You can either shelter your flowers from the rain or add a sprinkler system. Does this data[16] give you direction in what to do?

**Table 7-9. Correlation data.**

| rain | orchids | rain | orchids | rain | orchids |
|------|---------|------|---------|------|---------|
| 4.4 | 40 | 4.5 | 38 | 4.9 | 39 |
| 4.7 | 41 | 3.8 | 33 | 4.0 | 33 |
| 4.6 | 39 | 4.6 | 42 | 4.3 | 40 |
| 4.2 | 37 | 5.1 | 42 | 5.1 | 44 |
| 4.2 | 37 | 4.8 | 43 | 4.1 | 34 |
| 4.6 | 38 | 4.2 | 35 | 4.8 | 43 |
| 4.4 | 39 | 4.7 | 39 | 4.4 | 36 |
| 3.9 | 36 | 4.2 | 38 | 4.6 | 41 |
| 4.5 | 42 | 4.3 | 39 | 4.4 | 39 |
| 4.0 | 36 | 4.3 | 34 | 4.4 | 37 |

When plotted, Figure 7-12, you see an upward trend, but how well does the number of flowers correlate with the rainfall? Let's see.

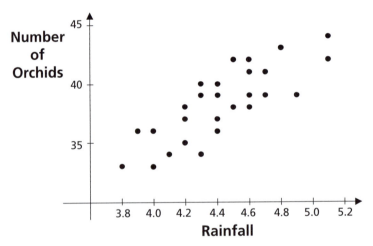

**Figure 7-12. Orchids vs. rainfall from Table 7-9. Correlation data.**

Before we introduce the **correlation coefficient**, let's normalize this data so we don't get tangled up in units and magnitudes that might skew our results. We'll normalize using properties of the data by letting

---

[16] This is not real data.

$$u_i = (x_i - \overline{x}) / S_x$$

and

$$v_i = (y_i - \overline{y}) / S_y$$

where

$\overline{x}$ = mean of the $x$, rainfall, data,

$\overline{y}$ = mean of the $y$, number of flowers, data,

$S_x$ = square root of the second moment of the $x$ data about $\overline{x}$, and

$S_y$ = square root of the second moment of the $y$ data about $\overline{y}$.

As expected, a plot of $v_i$ vs. $u_i$ (Figure 7-13) looks just like the plot of rainfall vs. flowers, except that we stripped it of its units and have centered the origin about the mean of both columns of data.

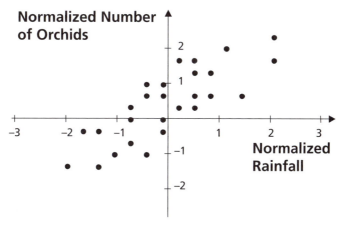

**Figure 7-13. Normalized orchids vs. rainfall data.**

Looking at our normalized graph, we see that if we multiply each pair of normalized data, $u_i \cdot v_i$, the product will be positive for data pairs in the first and third quadrants and negative in the second and fourth quadrants. So what do you think summing all of those products will tell us? If the number of orchids increases as more rain falls, we'll get a positive number for the sum of the products of $u_i \cdot v_i$, and we'll get a negative number if rain has the opposite effect on orchid growth. If the data is scattered all over the place, those products will sum to zero.

Now, before we list the value for the correlation, we have to take one more step. If we summed the products of the normalized $x$'s and $y$'s, (that is $\sum_i u_i \cdot v_i$) for that number of data points, we would have a number that expresses the valid correlation of that set of data. But if we added an equal number of equally scattered data points to that set, would our true correlation double? No, it would not.[17] So let's divide the sum of the products by the number of data points. This is the correlation coefficient $r$.

$$r = \frac{\sum_{i=1}^{n} u_i \cdot v_i}{n}$$

We don't have to go through the normalization steps for the $x$'s and $y$'s to get the $u$'s and $v$'s. We can go through all that arithmetic with one step and write the correlation coefficient, $r$, as

$$r = \frac{\sum_{i=1}^{n} (x_i - \bar{x})(y_i - \bar{y})}{n s_x s_y}$$

If the data is highly correlated, the sum of the products will approach 1 (or –1 if the number of flowers decreased monotonically as more rain fell). Our rainfall vs. flowers data show a correlation of 0.82.

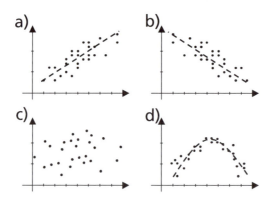

**Figure 7-14. Four sets of data showing correlation coefficients.**

Want to see some other correlation coefficients? Figure 7-14 shows four separate graphs of scattered data and their concomitant correlation coefficients. Note that graphs c and d have an $r$ factor of zero. Whereas graph c is obviously patterned with a twelve-gauge shotgun, graph d has a definite pattern to it, and it wouldn't take much tampering with the math to fit a $y(x) = ax^2$ equation to that data. What gives?

Is the correlation coefficient any good? Yes, but use it with care. For starters, it only applies to linear relationships. You saw what happened in Figure 7-14-d where $r$ was zero, and yet you saw a strong relationship between the pairs of data. You've also seen that too few sets of data can force a high coefficient. How much data is enough?

---

[17] Of course more data points make you more comfortable in stating conclusions. After all, two points will always produce a straight line, and unless that line is horizontal, the correlation will be either plus or minus one.

There is no easy answer to that one. And, a high correlation coefficient does not prove a relationship. We don't have a good example off hand, but we're sure that you've seen charlatans trying to prove all sorts of things with the phase of the moon and the alignments of the planets. So use the correlation coefficient, but use it with care.

# Curve Fitting[18]

We used the correlation coefficient to look at the linear relationship between sets of data. We will now quantify the best linear curve that describes the data. Let's use the rainfall vs. number of flowers data. You can take a straight edge, lay it upon the graph, and move it around to see where an equal number of points fall above and below your straight edge. Now you can draw your line and determine $b$ and $m$ to write the linear equation $y(x) = b + mx$. Or you can use the **method of least squares** to fit a curve (straight line) to the data.

Draw an imaginary line through your data and then mathematically measure how well you bisected all those data points. Then, try again to see if you do any better on the second try. Sounds messy, but our friend calculus has a better way for us to go.

First you'll write a general linear equation after moving the $y$-axis to the mean of the $x$-data. That is, you'll move the $y$-axis to $\bar{x}$ as

$$y = b + m(x - \bar{x}).$$

> Note that this $b$ is the intercept of the line on the $y$-axis that we placed at $x = \bar{x}$. The intercept on the original $y$-axis is $b_{old} = b - m\bar{x}$. The reason for introducing this confusing transformation will become obvious.

It's doubtful that these calculated values of $y$ will hit many of the $y$ data points; however, we want to minimize the amount of miss. Rather than summing the differences between the calculated $y(x_i)$s and the data point $y_i$s, let's square these differences since some of the $(y(x_i) - y_i)$s will be positive and some will be negative. Thus, the equation we want to minimize is

$$E(b, m) = \sum_{i=1}^{n} [y_i - y(x_i)]^2$$

$$E(b, m) = \sum_{i=1}^{n} [y_i - b - m(x_i - \bar{x})]^2$$

So $E$, the sum of the squared errors, is a function of only $b$ and $m$, and you know what to do to find the minimum of a function. You take the first derivative and set it to zero. $E(b, m)$ is a function of two variables; therefore, you must set each partial derivative to zero.

---

[18] This procedure is often called *regression*. This is a biological term for the return (i.e., regression) of a complex population to a less complex type in successive generations. We haven't studied much biology, but we've fit many a curve, so we're using *curve fitting*. You should know the other term, however.

Here it goes:

$$\frac{\delta E}{\delta b} = \sum_{i=1}^{n} \{-2[y_i - b - m(x_i - \overline{x})]\}$$

and

$$\frac{\delta E}{\delta m} = \sum_{i=1}^{n} 2[y_i - b - m(x_i - \overline{x})][-(x_i - \overline{x})].$$

Setting each of these derivatives to zero gives us

$$bn + m\sum_{i=1}^{n}(x_i - \overline{x}) = \sum_{i=1}^{n} y_i$$

and

$$b\sum(x_i - \overline{x}) + m\sum(x_i - \overline{x})^2 = \sum(x_i - \overline{x})y_i.$$

Note that

$$\sum_{i=1}^{n}(x_i - \overline{x}) = 0$$

hence

$$b = \overline{y}$$

and

$$m = \frac{\displaystyle\sum_{i=1}^{n}(x_i - \overline{x})y_i}{\displaystyle\sum_{i=1}^{n}(x_i - \overline{x})^2}$$

So the equation for our least squares curve fit (regression line) is

$$y(x) = \overline{y}_i + \left( \frac{\displaystyle\sum_{i=1}^{n}(x_i - \overline{x})y_i}{\displaystyle\sum_{i=1}^{n}(x_i - \overline{x})^2} \right)(x_i - \overline{x}).$$

We replot the graph of rainfall vs. the number of flowers again, but this time we show our calculated curve (well, a straight line) through the data in Figure 7-15.

Looking at our newly formulated line slicing through the data points helps verify the trend that in your garden rainfall promotes orchid growth. With the amount of scatter, however, you are probably not entirely satisfied. You would like to see more data. You should look at other factors such as cloudy days, temperature variations, and the amount of plant food. Curve fitting is only part of the story. Like correlation—handle with care.

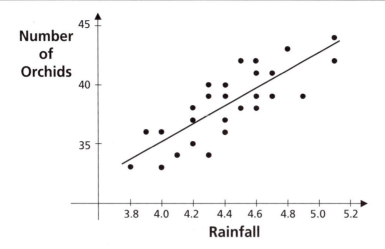

**Figure 7-15. Curve fit of orchids vs. rainfall data.**

And what about that parabolic looking curve in Figure 7-14-d? No, we are not going to fit a linear curve to it, for doing so would only produce a horizontal line. That's the same result as a linear curve fit to a set of data that has a linear correlation coefficient of zero. The good news is that you can curve fit using polynomials (polynomial regression some call it) of degree greater than one. The bad news is that we are not going to do it here—just remember that it can be done.

## How Do You Know if the Die Is Honest?— the $\chi^2$ Test

You roll the die and expect any number between one and six to show. The *one* shows. You roll again with the same expectation. The *one* shows. If you're a trusting person, you calculate the probability of that outcome and keep playing. The *one* comes up again. Now are you still trusting? If you are, we have a bridge we want to sell you.

Of course, it's possible to roll an honest die sixty times and get sixty *ones*, but not likely. The $\chi^2$ test, called chi-square, was developed to help us keep the gamblers honest. (If anyone knows where the name chi-square came from, please let us know.

Assume that we have the following results from sixty rolls of a die

**Table 7-10. Data for a chi-square test.**

| Face | 1 | 2 | 3 | 4 | 5 | 6 |
|---|---|---|---|---|---|---|
| Observed | 14 | 6 | 7 | 8 | 13 | 12 |
| Expected | 10 | 10 | 10 | 10 | 10 | 10 |

and let's use $\chi^2$ to help test the die's honesty.

$$\chi^2 = \sum_{i=1}^{n} \frac{\left(o_i - e_i\right)^2}{e_i}$$

where:

   $o_i$ is the observed outcome for the i[th] experiment,

   $e_i$ is the expected result for the i[th] experiment, and

   $n$ is the number of experiments.

For this set of data

$\chi^2 = 5.8$.

So is the die honest or not? We don't know. As we said before, honest dies can give unexpected results. Mothers in an upstate New York hospital once gave birth to sixteen boys without a girl birth. The probability of sixteen boys in row is 65,536 to 1, but it happened.

To expect dishonesty of the die, you would either have to compare the $\chi^2$ value for many (how many?) sets of sixty rolls, or you could calculate the probability, using the binomial method, of rolling that distribution.

## Confidence Interval

If you have measured a few thousand Texas longhorn's horns and were careful to choose your samples in a broad random manner, you might feel confident that you have a Gaussian distribution representing all Texas longhorn horns. Then you meet Cowboy Bob and Cowgirl Jane. He has one hundred longhorns with an average width of 1 meter, and she has nine with an average width of 1.1 meters. She claims that her cattle-raising methods are better than Bob's. He studied statistics at A&M and says that her sample is too small to tell. Can you help them settle their argument?

You remind Bob and teach Jane about the ***confidence interval***. With the confidence interval you assume that the true mean of each of their herds is distributed about their small sample measurement, $\bar{x}$, with a deviation of

$$\sigma_x = \frac{\sigma}{\sqrt{n}}$$

where:

   $\sigma$ is the standard deviation of Longhorn horns

and

   $n$ is the number in Bob's or Jane's sample.

If the standard deviation, as determined from many other measurements, is 0.35 meters, then deviation of the mean for an expected 68% ($\pm$ one standard deviation) of Bob's cattle is

$$\sigma_{Bob} = \frac{0.35}{\sqrt{100}} = 0.04 \text{ meters}$$ (Yes, we know that 0.35 divided by ten isn't 0.04. If this bothers you, re-read the section on measurements.)

and for Jane's

$$\sigma_{Jane} = \frac{0.35}{\sqrt{9}} = 0.12 \text{ meters}$$

Hence from this limited data you'd expect, to a 68% confidence level, that Bob's herd's horns would average from 0.96 to 1.04 meters and Jane's herd's horns to average from 0.98 to 1.22 meters. It's really like predicting a close election: it's too close to call.

## Probability and Statistics Examples

### The Rope Test

Let's assume that you must choose a rope supplier for your project and you want a breaking strength of greater that 90 newtons. You go to Ajax Rope Company and they offer you their 100-newton rope at $0.10 per foot, and you also talk to the Acme Company that wants $0.15 per foot for their 100-newton rope. Being a careful engineer you buy samples of each company's ropes, test them to their breaking points, and sure enough, the mean breaking point of each set is 100 newtons.

Is this the end of your quest? Do you pick the low bidder and continue with your project? Let's look at the data before making the final decision.

Table 7-11 and Table 7-12 are the frequency functions for the two sets of ropes.

### Table 7-11. Rope breaking tests from Acme Co.

| KG | Number |
|----|--------|
| 88 | 3 |
| 90 | 6 |
| 92 | 12 |
| 94 | 17 |
| 96 | 19 |
| 98 | 24 |
| 100 | 26 |
| 102 | 24 |
| 104 | 17 |
| 106 | 10 |
| 108 | 10 |

| | |
|---|---|
| 110 | 7 |
| 112 | 6 |
| 114 | 5 |
| 116 | 4 |

**Table 7-12. Rope Breaking Tests from Ajax Co.**

| newtons | Number |
|---------|--------|
| 70 | 1 |
| 72 | 2 |
| 74 | 0 |
| 76 | 3 |
| 78 | 5 |
| 80 | 5 |
| 82 | 6 |
| 84 | 6 |
| 86 | 7 |
| 88 | 9 |
| 90 | 0 |
| 92 | 9 |
| 94 | 10 |
| 96 | 9 |
| 98 | 10 |
| 100 | 12 |
| 102 | 11 |
| 104 | 11 |
| 106 | 10 |
| 108 | 10 |
| 110 | 9 |
| 112 | 8 |
| 114 | 9 |
| 116 | 8 |
| 118 | 4 |
| 120 | 6 |
| 122 | 5 |
| 124 | 2 |
| 126 | 2 |
| 128 | 1 |

You have tested 190 sections of rope from each company and after doing the arithmetic you are convinced that the average breaking strength is essentially the advertised 100 newtons. Now you calculate the standard deviation of the strength of the ropes from each supplier. Recall that the equation for the standard deviation is

**222**

$$\sigma = \sqrt{\sum_x f(x)*(\mu - x)^2 / n}$$

where:

$f(x)$ is the number ropes breaking at $x$ force,

$\mu$ is the average breaking strength,

and

$n$ is the total number of samples.

The data, Table 7-13, now tells a different story.

**Table 7-13. Average Breaking Strengths and the Deviation.**

|  | Ajax | Acme |
|---|---|---|
| $\mu$ | 100.6 | 100.6 |
| $\sigma$ | 13.0 | 6.4 |

After looking at the frequency functions for the ropes from both Ajax and Acme and calculating the mean and standard deviation of samples from both suppliers, you may decide to see how well the data fits a Gaussian distribution function. That is, you have $\mu$, the mean value, and $\sigma$, the standard deviation, so you put it into

$$f(x) = \frac{1}{\sigma\sqrt{2\pi}} e^{-\frac{1}{2}\left(\frac{x-\mu}{\sigma}\right)^2}.$$

The Gaussian distribution and the data for the two sample sets are shown in Figure 7-16.

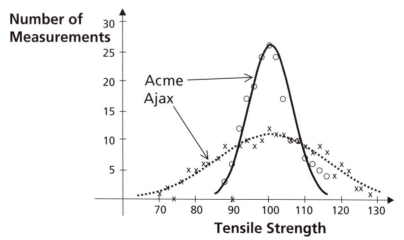

**Figure 7-16. Data for two sample sets and the Gaussian distributions calculated from the mean and standard deviation of the data.**

So, which rope do you choose? Do you take Ajax at $0.10 per foot and a standard deviation of 13 newtons, or do you spend the extra nickel for Acme with its standard deviation of 6.4 newtons? It all depends on your project. If you're using the rope to border your orchid beds, the extra money might be better spent on fertilizer. If the project is mountain climbing, and you still decide to go with Ajax, you had better put the savings into life insurance.

## Gaussian Distribution Example

The data in the previous example of rope-breaking tests was fictitious although we believe that it shows trends and points to decisions that engineers face in the real world. However, we want to work with some real physical data, and not only real data, but interesting data. We want to see if the length distribution of Texas Longhorns' horns approximates a Gaussian distribution.

Armed with a note pad, pencil, and tape measure we descended on the Fort Worth, Texas stockyard. We found a field of Longhorns, and sitting on the fence we called them to come over for their measurements. They didn't come. As very unselfish math book writers, not one of us wanted to hog the glory of being the one to go beyond the fence to make these measurements. We went across the street and ate BBQ Longhorn ribs instead.

While gnawing on the delicious lunch, we decided that the next wildest thing to the Longhorns was the Fort Worth weather. Though not as much fun, gathering the weather data was much easier: NOAA has done it for us. After looking at Fort Worth weather, we decided to compare it with the site of our future orchid ranch in Honolulu, Hawaii.

The frequency functions for the measured temperatures (seven measurements per day for one year) are shown in Table 7-14 and Table 7-15. Note that the wide range of temperatures prompted us to discretize the data into five-degree steps. Honolulu's temperatures were also discretized as the measurements were made to two decimal points, but we only used one-degree steps. (We hope that all of you metric folks will forgive the use of Fahrenheit temperature data.)

### Table 7-14. Fort Worth Temperature Data.

| Temp F, | Number |
|---------|--------|
| 17.5 | 2 |
| 22.5 | 9 |
| 27.5 | 19 |
| 32.5 | 46 |
| 37.5 | 78 |
| 42.5 | 140 |
| 47.5 | 194 |

| | |
|---|---|
| 52.5 | 185 |
| 57.5 | 227 |
| 62.5 | 216 |
| 67.5 | 263 |
| 72.5 | 257 |
| 77.5 | 254 |
| 82.5 | 283 |
| 87.5 | 278 |
| 92.5 | 100 |
| 97.5 | 4 |

**Table 7-15. Honolulu Temperature Data**

| Temp F, | Number |
|---|---|
| 65 | 1 |
| 66 | 4 |
| 67 | 8 |
| 68 | 18 |
| 69 | 29 |
| 70 | 51 |
| 71 | 65 |
| 72 | 104 |
| 73 | 182 |
| 74 | 253 |
| 75 | 238 |
| 76 | 254 |
| 77 | 228 |
| 78 | 232 |
| 79 | 272 |
| 80 | 270 |
| 81 | 196 |
| 82 | 115 |
| 83 | 32 |
| 84 | 3 |

Since we had all of this data, we started working with it while eating dessert at our next visit to the stockyards. We got

| | Ft. Worth | Honolulu |
|---|---|---|
| mean | 66.5 F | 76.7 F |
| deviation | 16.4 F | 3.4 F |

If you only look at the mean temperature, you might think that Fort Worth is the cooler place to hang out. Not so. Look at the deviation before you choose your sun hats and sweaters. The range of temperatures in Fort Worth that year was 17.3 to 96.6 °F while Honolulu only varied from 65.7 to 84.7 °F. So with the higher mean temperature, Honolulu was never as hot as Fort Worth.

The temperature distributions and the calculated Gaussian distributions are plotted in Figure 7-17. It's not a perfect fit, especially with the hotter Fort Worth temperatures, but it does show the usefulness of using the Gaussian equation to fit physical data.

More solved problems can be found on the accompanying CD-ROM.

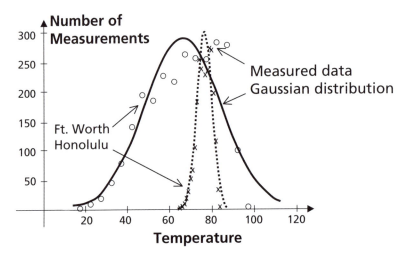

**Figure 7-17. Temperature Data for Fort Worth and Honolulu.**

# *Differential Equations*

*Though this be madness, yet there is a method in it.* — **William Shakespeare, *Hamlet***

We are now getting into the really heavy-duty stuff: differential equations. This is the ultimate language of the engineer and scientist.

"But wait a minute," you say. "We covered derivatives and differentials in the calculus chapter. The derivatives were written as equations, so I already know differential equations."

What you say is partially true. We covered derivatives, and you probably know all you need to know to understand what we'll present here.

In the calculus chapter we took Galileo's measurements for his falling cannon balls, fitted the data to a curve, and concluded that the force of gravity[1] caused the ball to fall with an acceleration of 9.8 meters/sec$^2$ which was determined from the second derivative of the position vs. time curve. That is,

$$g = \frac{d^2 x}{dt^2} \, .$$

If you knew $g$, and you knew that the ball started falling from rest and knew that no other forces[2] were acting on the ball, you then could integrate once and determine the ball's velocity as a function of time. If you wanted position as a function of time, you'd just integrate once more. As you may recall, the resulting equations are

$$v = \int g \, dt = gt + C_1$$

and

$$x = \int v \, dt = \frac{1}{2} g t^2 + C_1 t + C_2$$

---

[1]  Galileo almost discovered gravity. Read the history; it's fascinating. By the way, Galileo died the year that Newton was born—anyone for reincarnation?

[2]  Yes, that's a bunch of assumptions, but most book calculations are done in a perfect world with few, if any, extraneous perturbations. You'll need a computer and knowledge of numerical analysis, which we'll introduce in the chapter on computer math, to approach much of the real world.

where $C_1$ and $C_2$ are constants which we called $V_0$ and $D_0$.

That's all well and good (and those are differential equations).

Now, let's look at another force that can move things around. Let's look at the spring that stores[3] a force by compression. The force stored in our spring is

$$F_s = kx$$

where

$k$ is the spring constant and,

$x$ is the distance that the spring is compressed.[4]

Let's take that spring, fix one end to a rigid vertical wall, and fix the other end to a mass[5] sitting on a frictionless surface, as sketched in Figure 8-1. Compress the spring a distance $X$ and then release it. You intuitively know what will happen; the mass slides back and forth as the spring first goes back to its unperturbed position and keeps moving. When the spring reaches its normal length, it is exerting no force on the mass, but the moving mass is now a force pulling on the spring. The moving mass extends the spring until the spring has stretched as much as it was compressed. Motion ceases for a very brief moment, and the mass starts moving back toward the wall. This cyclical motion continues on and on.

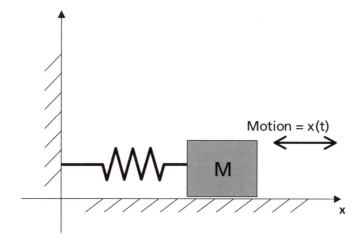

**Figure 8-1. Simple spring-mass system.**

---

[3] To all you physicists out there, we know that it's energy being stored, not force. Energy is force times the distance over which the force is acting.

[4] The spring stores the energy in both compression or extension. The only difference is the direction of initial movement when that stored energy accelerates a mass over a distance. We'll sort these signs out with a diagram.

[5] To keep life simple, we assume that the spring has no mass.

In real life, friction both within the spring and between the sliding surfaces and the viscosity of the mass moving through the air will slow the action. Each cycle of the spring-mass movement will be less than the last and eventually all action will cease. We'll soon get to that. For now let's write an equation for this frictionless motion.

We have two forces acting here. The force of the spring (Hooke's law) is

$$F_s = ks \,.$$

The force of the moving mass (Newton's second law) is

$$F_m = m \frac{d^2 x}{dt^2} \,.$$

These two forces are acting in opposite directions, so

$$m \frac{d^2 x}{dt^2} = -kx \,.$$

Written in its standard form, the equation of motion for this system is

$$\frac{d^2 x}{dt^2} + \frac{k}{m} x = 0 \,.$$

A quick look at our example shows that $x = 0$ when the spring is neither compressed nor extended. Does that mean the acceleration of the mass is also zero when it's in that position? Yes, it does. To see this, just consider the start of the first cycle. When the spring is first released, the force on the mass is maximum, and the acceleration (not velocity) is also maximum. The spring continues to push on the mass—though not as hard, hence with lesser acceleration. The velocity is still increasing—until the spring extends to its unperturbed length. At $x = 0$ the velocity of the mass is maximum, but it's no longer accelerating. The acceleration passes through zero and will become negative, slowing down the mass. If this argument reminds you of the maximums and minimums that you saw in calculus: good!

So what is $x$ as a function of time? Let's guess a solution and see if it works. (If guessing doesn't sound very elegant, we agree. But, this procedure works. And besides, it is an educated guess.) Let's guess that the solution is

$$x(t) = A \sin(at) \,,$$

and see if it works.

Since

$$\frac{d^2}{dt^2} A \sin(at) = -a^2 \, A \sin(at)$$

then the guessed solution seems to work if

$$a = \sqrt{\frac{k}{m}} \, .$$

Please note that we said, "seems to work" because we still haven't determined the value of $A$. Let's apply what else we know to see if we can solve for $A$. We know that we had compressed the spring an amount $x = X$ when we released the mass at $t = 0$. We also know that the mass was not moving at $t = 0$. That is, at $t = 0$, $v(t) = 0$. These facts are useful in solving the differential equation that describes the physical system. They are called the *initial conditions* of the equation. So, applying the initial conditions

$$x(0) = X$$

and

$$\frac{dx(0)}{dt} = 0$$

to

$$x(t) = A \sin \sqrt{\frac{k}{m}} t$$

and

$$\frac{dx(t)}{dt} = \sqrt{\frac{k}{m}} A \cos \sqrt{\frac{k}{m}} t$$

results in

$$A \cdot 0 = X \, ,$$

and

$$A\sqrt{\frac{k}{m}} = 0 \, .$$

This tells us that we don't have a useful equation. Our first guess solved the differential equation[6], but did not give us a physical solution when we applied all that we knew about the system.

Since the sine function almost worked, let's try her brother, the cosine function. Try

$$x(t) = B \cos bt \, .$$

---

[6] The solution $x(t) = 0$ solves the differential equation—stick it in and see for yourself. This is the trivial solution, and as the name implies, it won't tell us much about our physical world.

Differentiate twice and set the result equal to *k/m* times the function, and we have

$$-b^2 B \cos bt + \frac{k}{m} B \cos bt = 0 \,.$$

The cosine function is a realistic solution, including the initial conditions, if

$$b = \sqrt{\frac{k}{m}}$$

and

$$B = X.$$

Hence, the solution to our differential equation is

$$x(t) = X \cos \sqrt{\frac{k}{m}} t \,.$$

That was rather long and clumsy. Counting the trivial solution, we guessed three different solutions before we found the right one. You may wonder if there is a better way. There is, but it's still guesswork.

Remember that useful irrational number that we call *e*? It has the property that its derivative with respect to the variable in its exponent was just equal to itself. That is, if

$$f(x) = Ae^x$$

then

$$\frac{d}{dx} f(x) = Ae^x$$

and

$$\frac{d^n}{dt^n} = Ae^x \,.$$

Of course, if the exponent has a constant with the variable, we must use the composite function rule. That is, if

$$f(x) = Ae^{ax}$$

then

$$\frac{d}{dx} f(x) = aAe^{ax}$$

and

$$\frac{d^n}{dt^n} = a^n Ae^{ax} \,.$$

Now look at our example differential equation. It is a sum of our sought-after function and derivatives of that function. If our trial function can contain $e$, then we may have a simplified solution. Let's try it. Let our trial function be

$$x(t) = Ae^{at}$$

and then our differential equation will look like

$$a^2 Ae^{at} + \frac{k}{m} Ae^{at} = 0 ,$$

and our equation is reduced to an algebraic problem. We can factor out but not throw away our original function and solve

$$a^2 + \frac{k}{m} = 0$$

for $a$ which is

$$a = \pm \sqrt{-k \Big/ m}$$

or

$$a = \pm i \sqrt{\frac{k}{m}}$$

where $i$ is $\sqrt{-1}$.

So now we have the solution for our differential equation of

$$x(t) = Ae^{i\sqrt{\frac{k}{m}}t} + Be^{-i\sqrt{\frac{k}{m}}t} .$$

Where did $B$ come from?

For now $B$ is just an arbitrary constant like $A$. We got two solutions for the differential equation, and so far there is nothing to predicate a value of the constants in front of those functions—so let's give each one a different constant.

Before we try to make sense of what appears to be a messy solution, let's observe a few things.

1)  The solution as stated above is a sum of two separate functions. Each function, separately, is called a ***specific solution***.

2)  Either of the two specific solutions will solve the differential equation separately, or we can solve the differential equation with the sum. The sum is called the ***general solution***. This is the ***Principle of Superposition***. We saw the same thing when we used sine and cosine to guess the solution of the differential equation.

3) At this point, the coefficients, *A* and *B*, are arbitrary (*A* and *B* can even be zero, but then we'd have the trivial solution). Since *A* and *B* are arbitrary and the functions can be used separately, we can write the completed general solution as

$$x(t) = Ae^{i\sqrt{\frac{k}{m}}t} + Be^{-i\sqrt{\frac{k}{m}}t}$$

where *B* may or may not equal *A*. The values of *A* and *B* will be determined when we apply the initial conditions.

4) Our original guesses with the sine and cosine can be written as

$$x(t) = A\sin at + B\cos bt .$$

The algebra would have determined that $a = b = \sqrt{\frac{k}{m}}$, and the initial condition would have determined that $A = 0$ and $B = X$ as we had shown.

5) So far, we have related everything we've done with differential equations to a physical phenomena. As we've said, "Differential equations are the language of the engineer and scientist," but please be assured that differential equations also stand alone as a mathematical entity.

With all that said, let's try to make sense of our exponentials. For that, we pull in what seems like magic, the *Euler-de Moivre* formula, which states that

$$e^{iat} = \cos at + i\sin at .$$

If you say that it looks strange we can only agree with your casual observation, but the identity works. Just expand both sides in a *Maclaurin* expansion, and you'll see through the haze.

So now with the Euler-de Moivre formula and the fact that the coefficients *A* and *B* are arbitrary (they can even incorporate *i*) until the initial conditions are applied, we hope you agree that

$$x(t) = Ae^{i\sqrt{\frac{k}{m}}t} + Be^{-i\sqrt{\frac{k}{m}}t}$$

and

$$x(t) = A\sin\sqrt{\frac{k}{m}}t + B\cos\sqrt{\frac{k}{m}}t$$

are really the same solutions to our differential equation. The multiplying constants, *A* and *B*, will be different in the two equations, but *x(t)*, for any *t*, will have the same value regardless of which form we use.

Now that we've played around with that oscillating mass-spring system until you're sick of it, let's put it aside and look at some definitions. You'll soon tire of these definitions and be ready to come back to the mass-spring system (with more features applied).

1) The most general form of a ***linear differential equation*** is written as

$$f_n(t)\frac{d^n x(t)}{dt^n}+f_{n-1}(t)\frac{d^{n-1}x(t)}{dt^{n-1}}+...+f_2(t)\frac{dx(t)}{dt}+f_1(t)x(t)=h(t).$$

2) Should the function $x(t)$ or one of its derivatives be taken to a degree greater than one, then the differential equation is ***nonlinear***. That is,

$$t^2\frac{dx(t)}{dt}+\sin tx(t)=25$$

and our spring-mass differential equation are both linear. The differential equation

$$\left(\frac{dx(t)}{dt}\right)^2+x(t)=0$$

is not linear.

3) The general linear differential equation is of order $n$, the highest-order derivative in the equation. (Do not confuse *order* with *degree*.)

4) If $h(t)=0$, then the equation is said to be ***homogeneous***.

5) If $h(t)\neq 0$, then we have a ***nonhomogeneous differential equation***. The solution to a nonhomogeneous equation is obtained as though the equation were homogeneous, and then the solution is expanded to include the right-hand side.

   5a) The solution for the nonhomogeneous differential equation as though it were homogeneous is called the ***complementary function***.

   5b) The solution that considers only the right-hand side for *guessing* the solution is called the ***particular solution***. The ***general solution*** is the sum of the complementary function and the particular solution.

   Confusing? Yes, but we'll straighten it out.

6) Should you write the differential equation without the function in place, you would have what is called a ***linear differential operator***. The linear differential operator is often represented by the letter, $L$. That is,

$$f_n(t)\frac{d^n x(t)}{dt^n}+f_{n-1}(t)\frac{d^{n-1}x(t)}{dt^{n-1}}+...+f_2(t)\frac{dx(t)}{dt}+f_1(t)x(t)=h(t)$$

can be written as

$$Lx(t) = h(t)$$

where

$$L = f_n(t)\frac{d^n}{dt^n} + f_{n-1}(t)\frac{d^{n-1}}{dt^{n-1}} + \ldots + f_2(t)\frac{d}{dt} + f_1(t).$$

The concept of an operator is valid because the same differential equation may describe many different physical phenomena. The same equation might apply to a mechanical system in one case and an electronic circuit the next time. Units and initial conditions will change, but the general solution will remain. We won't do much with the linear differential operator, but you need to have heard of it.

7)  Linear differential equations in which all of the coefficient functions, that is all of the $f_i(t)$'s, are constants are called *linear differential equations with constant coefficients*. Yes, we realize that it sounds like a no-brainer definition, but a large group of important differential equations fall in this definition, and it is these equations that are solved by assuming the $e^{at}$ type of solutions.

8)  When you assume the exponential solution in a linear differential equation with constant coefficients and then factor out the exponential function, the algebraic polynomial remaining is the *characteristic equation* of your differential equation. For example $b^2 + \frac{k}{m} = 0$ is the characteristic equation for our simple spring-mass system.

That's enough definitions for a while. Let's move on.

## Complete and Independent Solutions

We have seen that our sample second-order differential equation had two functions for the general solution. That is,

$$x(t) = A\sin\sqrt{\frac{k}{m}}t + B\cos\sqrt{\frac{k}{m}}t$$

was the general solution subject to the initial conditions. (We could have expressed the solution in the exponential form but it is really the same thing.) You might wonder if these are the only solutions, in addition to the trivial solution, and if these are really two separate independent solutions. Realize that the number of functions in the general solution came from solving an algebraic polynomial. As we saw in algebra, a polynomial of degree $n$ will have $n$ roots. We expect the same thing even when the origin of the polynomial is a differential equation. So we expect two functions in the solution for a second-order differential equation such as our spring and mass, and we'd expect *n* functions for a differential equation of order *n*. Of course, if we have *n* specific solutions for a $n^{th}$ order differential equation, we'll have to have *n* initial conditions to solve for all of the coefficients of the solutions. And what do we do about repeating roots? There's another trick to handle that case, and we'll get to it later in this chapter.

To determine if you have independent[7] functions for your general solution, think about the conditions needed to solve a set of equations with many unknown variables. You needed *n* independent equations to solve for *n* unknowns. You then put those *n* equations into matrix form and solved for your unknowns. Part of that matrix operation involved calculating the value of the determinate of the coefficients of your *n* equations. If the determinant was equal to zero, then your equations were not independent.

So what does this have to do with differential equations? It's a close analogy. If you have a second-order differential equation, you solve it and get two functions with arbitrary coefficients. The coefficients are determined when you apply the initial conditions to the functions and the first derivatives of the functions. For example, with that old spring-mass system, we had the general solution of

$$A\sin\sqrt{\frac{k}{m}}t + B\cos\sqrt{\frac{k}{m}}t$$

which to save some ink we'll call

Au(t) + Bv(t).

To solve for *A* and *B* we needed two equations, which we got from the initial conditions of the general solution and its first derivative. In our earlier example, the first derivative was zero at *t* = 0, but that doesn't have to be the case. Either a different coordinate system or the choice of a different start time would have given different initial conditions[8], so we can write these equations, evaluated at their initial conditions, as

$$Au(t_i) + Bv(t_i) = C_1$$

and

$$A\frac{du(t_i)}{dt} + B\frac{dv(t_i)}{dt} = C_2$$

---

[7] Please remember what we mean by independent functions. To be independent, one or more functions can not be written as a combination of the other functions in the set. For example the functions $\sin^2 t$, 1, and $\cos^2 t$ are not independent since $\sin^2 t + \cos^2 t = 1$.

[8] These arbitrarily chosen times or coordinates would not have changed the physics of our system. The mass would oscillate just the same. Only the way we describe the system would change.

where

$C_1$ is the value of the general solution at $t_i$

and

$C_2$ is the value of the first derivative of the general solution at $t_i$.

We have two equations and two unknowns. You know $u(t_i)$ and $v(t_i)$ as well as the derivatives of $u$ and $v$ at $t_i$. You also know the initial conditions, $C_1$ and $C_2$. To solve for the unknowns, $A$ and $B$, the determinant of

$$\begin{vmatrix} u(t) & v(t) \\ du/dt & dv/dt \end{vmatrix}$$

must not vanish (that is, not equal to zero) for any value of $t$ over the interval of interest.

"Why must it not vanish over the whole interval of interest?" you ask. "Don't we calculate the arbitrary constants at only one value of the independent variable? So, what if the determinant doesn't vanish at that one point, why isn't that good enough?"

Good question, but remember that we can't specify $t_i$, the time of those set conditions. We are free to make our measurements whenever we want to and we would still want the differential equation to be valid.

Needless to say, if we had an $n^{th}$ order differential equation, we would have an $n$-by-$n$ matrix whose determinant must not vanish.

This determinant of the $n$-functions and the $n$-derivatives of the general solution of your $n$-order differential equation is called the ***Wronskian***, named for the Polish mathematician, G. Wronski, who came up with it.

## A Quick Review

Let's go back to the spring-mass system. Before we add some frills to it, let's review what we've covered.

We had a homogeneous, linear, second-order differential equation with constant coefficients.

$$\frac{d^2 x(t)}{dt^2} + \frac{k}{m} x(t) = 0$$

Because it had those properties, we assumed a solution in the form

$$x(t) = A e^{at}$$

which when put into our differential equation gave us the characteristic equation

$$a^2 + \frac{k}{m} = 0$$

which has

$$x(t) = Ae^{i\omega t} + Be^{-i\omega t}$$

for the general solution which, thanks to Euler and de Moivre, we write as

$$x(t) = A\sin\omega t + B\cos\omega t$$

where

$$\omega = \sqrt{\frac{k}{m}}$$

> [And of course, the *A* and *B* are not the same for the two different ways of writing the general solution. Remember, *A* and *B* are arbitrary values until we apply the initial conditions.]

We then ask our friend Wronski if these are independent solutions, and he puts them into the Wronskian and gets

$$\begin{vmatrix} \sin\omega t & \cos\omega t \\ \omega\cos\omega t & -\omega\sin\omega t \end{vmatrix} = -\omega.$$

The solutions are independent unless $\omega = 0$, in which case we wouldn't have had any action anyway.

To determine *A* and *B*, we turn to the two initial conditions governing the action of this system:

$$x(0) = X$$

and

$$\frac{dx(0)}{dt} = 0$$

where *X* was the initial displacement of the spring.

This produces the equations

$$A * 0 + B = X$$

and

$$\omega A \cdot 1 - \omega B \cdot 0 = 0$$

The solution, with initial conditions, is

$$x(t) = X\cos\omega t.$$

# Spring-Mass System with Damping

We all know that our spring will not oscillate the mass forever, so let's get real. We'll put a damper on the system. This damper[9] could be like an automobile shock absorber or a sail that pushes through the air. In either case, the damper exerts no force on the mass unless it's moving, and then the force it exerts is proportional to the mass's velocity. The faster the mass moves, the greater the resisting force. Mathematically, the damper looks like

$$F_d = -r \frac{dx(t)}{dt}$$

where $r$ is a constant describing the damper, and our new differential equation is

$$\frac{d^2 x(t)}{dt^2} + \frac{r}{m} \frac{dx(t)}{dt} + \omega^2 x(t) = 0$$

where $\omega = \sqrt{k/m}$ as in the undamped system.

Following the same procedure with exponential function $x(t) = A e^{at}$ as our guessed solution, we get the characteristic equation

$$a^2 + a \frac{r}{m} + \omega^2 = 0$$

which tells us that

$$a = -\frac{r}{2m} \pm \frac{\sqrt{\left(\frac{r}{m}\right)^2 - 4\omega^2}}{2}.$$

Let $b = r/m$ and write the polynomial solution as

$$a = -\frac{b}{2} \pm \sqrt{\left(\frac{b}{2}\right)^2 - \omega^2}.$$

This gives us two functions for this second-order differential equation as you'd expect, but now we see three different forms that the two functions can take depending on the relative values of b and $\omega$. The three fundamentally different types of solutions depend on whether the value under the radical (square root) is negative, positive, or equal to zero.

Case 1. $\quad \omega^2 > \left(\frac{b}{2}\right)^2.$

---

[9] The term *dashpot* is often used for *damper*.

Here the value under the radical is negative, and we have an imaginary function as we had with the undamped case. The difference is that we have a complex conjugate[10] as the exponents. That is

$$x(t) = Ae^{-(\frac{b}{2}-i\xi)t} + Be^{-(\frac{b}{2}+i\xi)t}$$

where $\xi = \sqrt{\left|\left(\frac{b}{2}\right)^2 - \omega^2\right|}$

and, as before, the solution can be written as

$$x(t) = e^{-\frac{b}{2}t}\left(A\sin\xi t + B\cos\xi t\right).$$

> Before we go any further, please note that if the damping coefficient reduces to zero, the exponential in front of our familiar sine and cosine reduces to unity and our new frequency, $\xi$, reduces to $\omega$ just like the earlier example.

Again we have an oscillating mass on the end of a spring, but the mass won't mathematically vibrate[11] forever. The oscillations will die out as $e^{-\frac{b}{2}t}$ shown in Figure 8-2.

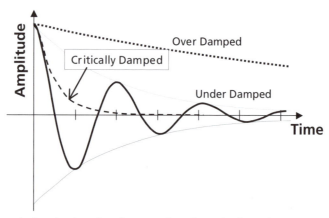

This is set up to look at the slamming of a screen door. The under-damped, case 1, swings back and forth while the air slowly stops the door. The over-damped has a damper mechanism that is set too tightly. The critically damped allows the door to return to the closed position as quickly as possible.

**Figure 8-2. Damped spring-mass system for the three levels of damping.**

---

[10] Our old solution, $a = \pm i\omega$, was also a complex conjugate with $b = 0$.

[11] Of course the equation $e^{-at}$ is never zero for finite $t$. It says that the oscillations will go on forever. We know that they don't. They'll cease long before the spring rusts away. Does the math fail the reality test? No! We just have to be realistic about the interval over which to expect realistic mathematical solutions for physical systems.

We apply the initial conditions as before, and solve for $A$ and $B$.

So

$$x(t) = Xe^{-\frac{b}{2}t}(\frac{b}{2\xi}\sin\xi t + \cos\xi),$$

which reduces to the undamped solution if $b = 0$.

An aside: We sometimes want to plot the derivative of the solution against the solution, and that's velocity plotted against position in this case. The plot made in $x(t)$, $dx/dt$ coordinates is called ***phase space***.

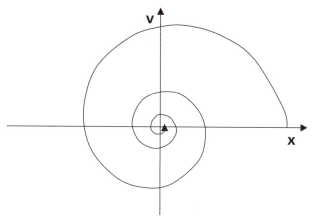

**Figure 8-3. Phase space plot of position and velocity
of a damped spring-mass system.**

The curve in phase space for the damped system is shown in Figure 8-3. As you can surmise from looking at the equation, it spirals into the origin of the coordinates. The undamped system would have produced a circle whose diameter would remain constant as determined only by the spring constant, the mass, and the initial displacement.

Case 2. $\omega^2 < \left(\frac{b}{2}\right)^2$. The over-damped case.

The value under the radical in the solution for $a$ is positive, so we have no imaginary number in the exponent, and we have no sine or cosine functions. With no sine or cosine functions, we have no oscillations. The spring and mass will move, but they will not hop about. Do you see the analogy with your automobile shock absorber? The worn-out shocks whose $r$ coefficient has decreased lets the car bounce. You'll get the same effect in an overloaded car, because increasing mass has the same effect as decreasing the damping coefficient.

Let's do the math.

The general solution looks as it did for the oscillating case except that the exponent has no imaginary component (which is a very big difference). It is

$$x(t) = Ae^{-(\frac{b}{2}-\xi)t} + Be^{-(\frac{b}{2}+\xi)t}.$$

The general solution is still subject to the same initial conditions, which form the following two equations:

$A + B = X$ , for the initial condition that $x(0) = X$

and

$$\left(\frac{b}{2} - \xi\right)A + \left(\frac{b}{2} + \xi\right)B = 0, \text{ for the initial condition that } \frac{dx(0)}{dt} = 0.$$

And the solution[12] is

$$x(t) = \frac{1}{2}Xe^{-\frac{b}{2}t}\left\{\frac{b}{2\xi}\left(e^{\xi t} - e^{-\xi t}\right) + \left(e^{\xi t} + e^{-\xi t}\right)\right\}.$$

*Case 3.* $\omega^2 = \left(\frac{b}{2}\right)^2$ The critically damped case.

In this case the radical is not only not imaginary. It doesn't exist. At first it seems that we have only one solution for this second order differential equation. How do we get that second solution which we hope you're convinced must exist? You pull another trick, that's how.

If we had a polynomial with repeating roots, such as $t^2 - 4t + 4 = 0$, which has a single root at $x = 2$, that was OK. But the general solution isn't

$$x(t) = Ae^{-\frac{b}{2}t} + Be^{-\frac{b}{2}t}.$$

These two are glaringly dependent. If you doubt us, stick them in the Wronskian.

We will tell you, without even a hand-waving proof, how to handle solutions with multiple roots in the characteristic equation. You pick the new solution

$$x(t) = Ae^{-\frac{b}{2}t} + Bte^{-\frac{b}{2}t}.$$

Sounds weird, but stick it in the differential equation and see for yourself that it works. You might also try it with the Wronskian to see that it passes the test.

---

[12] Some folks would want to stick hyperbolic functions into solutions such as this one. We won't mess with these functions other than to tell you what they are. The hyperbolic sine and cosine are defined as $\sinh(t) = \frac{1}{2}\left(e^t - e^{-t}\right)$ and $\cosh(t) = \frac{1}{2}\left(e^t + e^{-t}\right)$. You'll use these functions if you design suspension cables, among other things.

Assuming that we have the same initial conditions, we solve for $A$ and $B$ as

$$A + B \cdot 0 = X$$

and

$$-\frac{b}{2}A + B = 0$$

which leaves us with

$$x(t) = X\left(1 + \frac{b}{2} \cdot t\right)e^{-\frac{b}{2}t}.$$

A critically damped curve is also shown in Figure 8-2. The term *critically damped* comes from the fact that this is the least possible level of damping before the system oscillates. If you lower the damping coefficient by any amount, the system will oscillate—perhaps not for long, but it will oscillate.

## Multiple Roots to the Characteristic Equation

We just had an example of two equal roots, and we determined the general equation by multiplying one of the functions by the variable.

If we want to solve the differential equation

$$\frac{d^2 f(x)}{dx^2} - 4\frac{df(x)}{dx} + 4 = 0, \quad \text{(note that we're using } x \text{ as the variable, not the}$$

function as with the spring-mass system) we use the normal approach to solution as before with

$$f(x) = Ae^{at} + Be^{bt}$$

which gives us the characteristic equation

$$a^2 - 4a + 4 = 0.$$

This equation has multiple roots at

$$a = 2.$$

So the general solution (using our multiple root trick) is

$$f(x) = Ae^{2x} + B \cdot te^{2x}.$$

If we had

$$\frac{d^3 f(x)}{dx^3} - 6\frac{d^2 f(x)}{dx^2} + 12\frac{df(x)}{dx} - 8f(x) = 0,$$

we'd have triple roots at $x = 2$, and our general solution would be

$$f(x) = Ae^{2t} + B \cdot te^{2t} + C \cdot t^2 e^{2t}.$$

Higher-order roots would just continue to follow this pattern.

## Forced Oscillations

So far the actions of our spring-mass-damper system were only determined by the characteristics of the system itself and by the initial conditions that we had imposed upon it. We are now going to continue to impose a perturbation on the system that is more than just an initial condition. We are going to hit it with the oscillating force

$$F_f(t) = F \cos \vartheta t .$$

(We don't have to hit this mechanical system with only trig functions. We could use almost anything[13]. And please note, $F$ is not an arbitrary constant to be determined. $F$ is the amplitude of our imposed forcing function.)

The sum of our forces, hence the differential equation, is now

$$\frac{d^2 x(t)}{dt^2} + \frac{r}{m} \frac{dx(t)}{dt} + \omega^2 x(t) = F \cos \vartheta t .$$

You want to solve for $x(t)$.

It's easier than you might think.

If you remember our earlier definitions, you'll recognize this as a nonhomogeneous differential equation (assuming $F \neq 0$).

The first step in solving a nonhomogeneous differential equation is to solve it as though it were a homogeneous equation. After determining that solution, called the *complementary function*, you'll form a trial solution using the right-hand side of the differential equation and operate on it as though it were the only solution. This gives you the *particular solution*. The complementary function added to the particular solution is the *general solution*. Arbitrary constants in the complementary function are determined by applying the initial conditions to the general solution. Let's look at a couple of simple examples before we plow forward with the forced oscillations.

Look at the nonhomogeneous differential equation

$$\frac{d^2 f(x)}{dx^2} + f(x) = 3 + 4x \quad \{\text{And note that we're using } f(x), \text{ not } x(t).\}$$

You can easily find the solution for the homogenous part of this equation:

$$f_h(x) = A \sin x + B \cos x .$$

---

[13] Refer to the section in calculus on Fourier analysis to see that trig functions are not bad examples to use here.

To determine the particular solution we'll try a function of the form of the right-hand side of the differential equation. The right-hand side $(3 + 4x)$ is a linear function, so we'll try

$$f_p(x) = a + bx,$$

as the guess for the particular solution which we'll put into the nonhomogeneous equation as

$$\frac{d^2}{dx^2}(a + bx) + (a + bx) = 3 + 4x.$$

The coefficients, $a$ and $b$, of our particular solution[14] are solved:

$$f_p(x) = 3 + 4x.$$

The general solution, $f(x) = f_h(x) + f_p(x)$, is then

$$f(x) = A\sin x + B\cos x + 3 + 4x.$$

Apply your initial conditions to the general solution to determine A and B.

OK, that one was too easy. We'll do another one that's a bit more interesting. Solve:

$$\frac{d^2 f(x)}{dx^2} + 2\frac{df(x)}{dx} = 3 + 4x$$

The polynomial for the homogeneous part of this differential equation has roots at zero and –2. Hence,

$$f_h(x) = A + Be^{-2x},$$

which was easy enough.

Even though the right-hand side of the nonhomogeneous equation is a linear function as in our first example, we can't use it here. Why?

Reason #1. We cannot have a constant in $f_p(x)$ since a constant is a solution for the homogeneous equation. What difference does that make? Since a constant, as a solution, makes the homogeneous equation equal to zero, then a constant in the particular solution would also make the differential equation equal to zero and not equal to the finite, non-zero, value on the right hand side.

Reason #2. If we only have a constant and a linear term for the trial solution, the function will reduce to a constant when differentiated. A constant as the only solution on the left-hand side will not produce the given right-hand side result.

---

[14] Please note that the particular solution is NOT a sum of solutions as with the homogeneous differential equations. That is, the 3 and the $4x$ do not solve the differential equation separately—you must use the sum to solve the equation.

So let's try

$$f_p(x) = ax + bx^2$$

as our trial solution.

Put this trial into the differential equation, and we get

$$2b + 2a + 4bx = 3 + 4x.$$

Equating powers of $x$, we have

$$f_p(x) = \tfrac{1}{2}x + x^2.$$

The complete, general solution is

$$f(x) = A + Be^{-2x} + \tfrac{1}{2}x + x^2 \quad \text{(since we didn't state initial conditions, we can't}$$
solve for $A$ and $B$).

Now, let's return to our forced oscillation differential equation,

$$\frac{d^2x(t)}{dt^2} + \frac{r}{m}\frac{dx(t)}{dt} + \omega^2 x(t) = F\cos\vartheta t.$$

We've solved the homogeneous equation, so assuming that $\omega^2 > \left(\dfrac{b}{2}\right)^2$,
that solution is

$$x_h(t) = e^{-\frac{b}{2}t}\left(A\sin\xi t + B\cos\xi t\right)$$

where $\xi^2 = \left|\left(\dfrac{b}{2}\right)^2 - \omega^2\right|$.

Following the rule we just covered in picking trial solutions for the particular solution, we'll pick

$$x_p(t) = a\sin\vartheta t + b\cos\vartheta t$$

as our trial solution and put it into the non-homogeneous equation,

$$\frac{d^2(a\sin\vartheta t + b\cos\vartheta t)}{dt^2} + \frac{r}{m}\frac{d(a\sin\vartheta t + b\cos\vartheta t)}{dt} + \omega^2(a\sin\vartheta t + b\cos\vartheta t) = F\cos\vartheta t$$

and turn the crank.

The result is

$$x_p(t) = F\frac{\frac{r}{m}\vartheta}{\left(\omega^2 - \vartheta\right)^2 + (\frac{r}{m}\vartheta)^2}\sin\vartheta t + F\frac{\left(\omega^2 - \vartheta^2\right)}{\left(\omega^2 - \vartheta^2\right)^2 + (\frac{r}{m}\vartheta)^2}\cos\vartheta t.$$

The complete, general solution subject to the initial conditions is

$$x(t) = x_h(t) + x_p(t),$$

which we will neither solve for $A$ and $B$ nor write out.

The complete solution looks messy, but there's a lot of information hidden in that haystack.

1) Note that the homogeneous equation exponentially dies out with time. However, the particular solution has sine and cosine functions in it which do not decrease in amplitude as time increases. In this example, you can say that the homogeneous solution gives the ***transient results***[15] whereas the particular solution gives the ***steady-state results***.

2) If there is no damping, that is $r = 0$, then the sine term on the particular solution disappears[16]. Essentially, the damping puts a phase lag in the system to the response of the external perturbation.

3) An interesting case occurs when the perturbing force gets close to the natural frequency, $\omega = \sqrt{k/m}$. To make it simple, let's drop the damping term and set initial conditions ($x(0) = 0$ *and* $\dfrac{dx(0)}{dt} = 0$) such that the solution is

$$x(t) = \frac{F}{\left(\omega^2 - \vartheta^2\right)}\left(\cos\vartheta t - \cos\omega t\right).$$

Using your trig identities, you can write the general solution as

$$x(t) = \frac{2F}{\left(\omega^2 - \vartheta^2\right)}\left(\sin\{\tfrac{1}{2}(\omega+\vartheta)t\}\sin\{\tfrac{1}{2}(\omega-\vartheta)t\}\right).$$

This function is the basis of AM radio[17] seen by the lower-frequency amplitude modulating the higher frequency (see Figure 8-4). This is sometimes called the phenomena of beats.

---

[15] Have you ever noticed the lights dim when your refrigerator starts up? If so, you've observed the transient effect of the compressor coming up to speed. It takes more energy to get motors and compressors started than it takes to keep them running. Once the compressor is running at its desired rotation rate, the power drain reduces and the lights brighten once more. You'll see much more of transient vs. steady-state in advanced engineering courses.

[16] We expect that the cosine term of the particular solution would disappear if the forcing function was a sine wave instead of a cosine wave. If you work it out, let us know; we're just too lazy to do it.

[17] Of course, the radio transmitter is using coils and capacitors and not springs and masses. Remember, these equations are useful in lots of different phenomena.

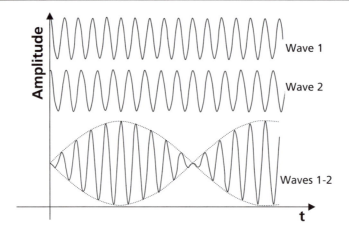

**Figure 8-4. The beat of two sine waves that are close to the same frequency.**

4)  As the frequency of the perturbing force gets close to the natural frequency, $\omega = \sqrt{k/m}$ , the amplitude of the response increases. If the damping term happens to be vanishing small, the response becomes exceedingly large. This is called resonance and is the phenomena that causes bridges to fall in the wind, crystal glasses to break from a singer's extended note and, maybe, walls to fall at the sound of a trumpet.

## Direct Integration

Though we alluded to it in the introduction of this chapter, we haven't said much about integrating the derivatives in these equations. And with good reason, we might add, because typically we can't. There are a few instances where we can integrate and get our solution. We'll show a couple of them before we move on to some fancy differential equations.

For the first example, we go back to the undamped, free oscillation equation

$$\frac{d^2 x(t)}{dt^2} + \frac{k}{m} x(t) = 0 .$$

The second derivative is the derivative of the first derivative, hence,

$$\frac{d^2 x(t)}{dt^2} = \frac{d}{dt} \frac{dx(t)}{dt} ,$$

The first derivative of position is velocity, so

$$\frac{d^2 x(t)}{dt^2} = \frac{dv(t)}{dt} .$$

Using the *chain rule*, we can write the second derivative as

$$\frac{d^2x(t)}{dt^2} = \frac{dv(t)}{dt}$$
$$= \frac{dv(t)}{dx(t)}\frac{dx(t)}{dt}.$$
$$= v(t)\frac{dv(t)}{dx(t)}$$

Put this back into our differential equation, and we have

$$v(t)\frac{dv(t)}{dx(t)} + \frac{k}{m}x(t) = 0,$$

which we rearrange as

$$v(t)dv(t) + \frac{k}{m}x(t)dx(t) = 0.$$

We can now integrate to get

$$\tfrac{1}{2}mv^2(t) + \tfrac{1}{2}kx^2(t) = C$$

where $C$ is a constant.

We have just proven the law of the conservation of energy because the constant, $C$, is the energy of the system. You put potential energy, $\frac{1}{2}kX^2$, into the system by compressing the spring the distance $X$. Although energy leaves the uncoiling spring, it goes into the kinetic energy, $\frac{1}{2}mv^2$, of the moving mass. The mass slows up, losing energy, by extending or compressing—depending on the direction of travel—the spring.

Were a damper in place we couldn't have solved the equation this way, but the conservation of energy law still holds. With a damper, some of the energy would be dissipated as heat, and the mass would slow down and the spring would extend (and compress) less on each cycle.

Another integratable differential equation is found in the analysis of charging and discharging capacitors, and in microbe growth or decay. We are talking about the phenomena whereby the larger a quantity is, the greater the change in that quantity. That mouthful is better stated as

$$\frac{df(x)}{dx} = kf(x)$$

where $k$ is a physical constant of the system.

We can rewrite this equation as

$$\frac{df(x)}{f(x)} = k\,dx.$$

When we integrate both sides we get

$$\ln\{f(x)\} = kx + c$$

where $c$ is a constant.

As we know $e^{\ln(y)} = y$ for a positive, nonzero value of $y$; therefore,

$$f(x) = ce^{kx}.$$

Whether this function is giving or taking away depends on the sign of $k$. Our arbitrary constant is $c$. If we have the value of $f(x_i)$ for some known $x_i$, we can completely solve the differential equation.

Needless to say, we'd have gotten the same solution had we guessed a solution in the form

$$f(x) = Ae^{ax}.$$

## Boundary Value Problems

The differential equations we've played with so far relied on knowing the value of the function and its derivative (if it were second order) at one specific value of the independent variable. There is another large class of physical problems described by differential equations where you don't know the values of the function and its derivative at a single value of the variable, but you know the value of the function at different values of the independent variable. Typically you know these values at the boundaries, so we call them **boundary value problems**.

An illustrative example of this class of problems is found in analyzing the Euler column. An Euler column is where a loaded column is pinned at one end and loaded and pinned at the other end as shown in Figure 8-5. It is known to remain straight until a critical load is reached.

Without getting into the mechanical engineering of buckling columns, we'll simply state the governing equation in which the shape is described by

$$\frac{d^2y(x)}{dx^2} + k^2 y(x) = 0$$

and

$$k^2 = PK$$

where

$P$ = equals the load

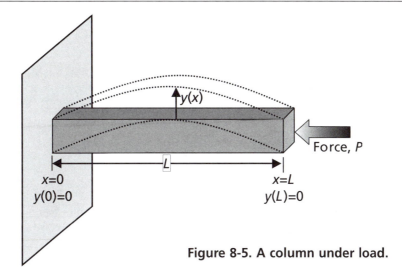

**Figure 8-5. A column under load.**

and

$K$ = materials properties.

Note that $k$ is an independent variable. You are increasing its value and watching the column bend.

Time isn't important[18] for this problem, but we do know that the ends do not move. That is the column at $x = 0$ and $x = l$ remains fixed. Hence,

$y(0) = 0$

and

$y(l) = 0$

where $l$ is the length of the column.

The differential equation looks like the equation for the undamped spring-mass system, and it has the same general solution, which is,

$y(x) = A\sin kx + B\cos kx$.

We don't have initial conditions to help solve for $A$ and $B$, so we'll use the boundary conditions.

$y(0) = B = 0$

---

[18] There is another phenomena that could come into play here. It's called *creep*, where the deformation of a loaded structure depends on time under that load. We'll leave that one for a more advanced book.

and

$$y(l) = A\sin kl = 0.$$

You have the trivial solution of *A=B=0* which is correct before buckling occurs, but that's not very interesting. The other solutions occur when

$$\sin kl = 0$$

which is when

$$k = n\frac{\pi}{l}.$$

So the first critical load, where the column first deflects, occurs when the load

$$P = \frac{\pi}{Kl}.$$

(Note that *n=0* is the same as the trivial solution of no bending.)

Other deflections will occur when *P* increases to

$$P = \frac{2\pi}{Kl}$$

and so on.

## Series Solutions to Differential Equations

There is another branch of differential equations that cannot be solved by the aforementioned means, but can be solved by assuming a power series as the solution. We will use a differential equation that can be solved by easier methods for our demonstration.

$$\frac{d^2 f(x)}{dx^2} + f(x) = 0$$

which, solved in the standard fashion, gives

$$f(x) = A\sin x + B\cos x.$$

Instead of using this easy method, we're going to assume a solution in the form

$$f(x) = A_0 + A_1 x + A_2 x^2 + \cdots + A_n x^n + \cdots$$

or using summation notation

$$f(x) = \sum_{i=0}^{\infty} A_i x^i.$$

"Hold on," you say. "I've learned that a second-order differential equation can have only two functions in its general solution. We just calculated those two, and now you say that we have an infinity of solutions."

And you are right, there are only two solutions. Just play the game and see what happens.

After you put the proposed series solution into the differential equation, you have

$$2A_2 + 6A_3x + 12A_4x^2 + \ldots + A_0 + A_1x + A_2x^2 + A_3x^3 + A_4x^4 + \ldots = 0.$$

Collecting like powers of $x$, we have

$$(2A_2 + A_0) + (6A_3 + A_1)x + (12A_4 + A_2)x^2 + \ldots = 0$$

The only way that this messy sum can equal zero, as demanded by the homogeneous equation, is for the coefficient of each power of $x$ to equal zero. Hence

$$2A_2 = -A_0, \qquad 6A_3 = -A_1, \qquad 12A_4 = -A_2, \qquad 20A_5 = -A_3 \text{ and so on.}$$

It's still a mess, but notice the recursion,

$$A_2 = -\frac{1}{2}A_0, \quad A_3 = -\frac{1}{6}A_1, \quad A_4 = -\frac{1}{12}A_2 = +\frac{1}{24}A_0, \quad A_5 = -\frac{1}{20}A_3 = +\frac{1}{120}A_1, \ldots$$

We can write the solution as two sums of the independent variable with only two arbitrary coefficients. Are you feeling better yet?

$$f(x) = A_0(1 - \frac{1}{2}x^2 + \frac{1}{24}x^4 - \ldots) + A_1(x - \frac{1}{6}x^3 + \frac{1}{120}x^5 - \ldots)$$

or if we use factorials in the equation, the solution is written as

$$f(x) = A_0(1 - \frac{1}{2!}x^2 + \frac{1}{4!}x^4 - \ldots) + A_1(x - \frac{1}{3!}x^3 + \frac{1}{5!}x^5 - \ldots)$$

Since each of these two series is independent of the two coefficients, $A_0$ and $A_1$, and each is linearly independent of the other, we can write the solution as

$$f(x) = A_0 u(x) + A_1 v(x).$$

We clearly see that the series method produced two and only two independent solutions to the second order differential equation. But you know from earlier sections in this chapter that the general solution to this differential equation is a sine plus a cosine. Now what you must do is refer all the way back to the section on series in calculus where you used Maclaurin's expansion to equate the sine function to the series shown as $v(x)$ as it's written above. Your assignment, now, is to use Maclaurin's, expansion on the cosine function and see if it equates to $u(x)$.

# Examples

## The Total Differential

We often have to work with a function of more than one variable, and we may need to know how that function changes when more than one of those variables are dancing about. Let's consider such a fictitious function $f(x,y,z)$, and see what we can do with it.

If we differentiate $f(x,y,z)$ with respect to $x$, we see how $f(x,y,z)$ changes with that one variable. That is

$$\frac{\partial f(x,y,z)}{\partial x} = \frac{f(x+\Delta x,y,z)-f(x,y,z)}{\Delta x}$$

as $\Delta x$ gets very, very small. Similar equations hold for $y$ and $z$.

If we want to see how $f(x,y,z)$ changes with respect to small changes with all three variables, we use the total differential, which is

$$df(x,y,z) = \frac{\partial f}{\partial x}dx + \frac{\partial f}{\partial y}dy + \frac{\partial f}{\partial z}dz\ .$$

(Don't worry about the use of $\partial$ and $d$ at different places. The use of $\partial$ just means that we are only looking at one of many variables: this is called a *partial derivative*.)

Of course we must keep $dx$, $dy$ and $dz$ small since what we are really doing is pretending that we can model the derivative of each of the variables (at $x$, $y$, $z$) as a linear function. You learned all about this linear approximation back in calculus. There we let the $dx$s get exceedingly small. Here you want them large enough to give meaningful changes in the function. Proceed with care.

As a numerical example look at

$$f(x,y) = 2xy - y^2\ ,$$

and ask how the function changes at $(0,0)$ and at $(8,10)$ where each time $dx = 0.1$ and $dy = 0.1$.

The general equation for the total differential at any point is

$$df(x,y) = \frac{\partial f}{\partial x}dx + \frac{\partial f}{\partial y}dy$$

$$= \frac{\partial(2xy-y^2)}{\partial x}dx + \frac{\partial(2xy-y^2)}{\partial y}dy\ .$$

$$= 2y\,dx + 2(x-y)dy$$

And at $(0,0)$ both of the partial derivatives are zero, so the total differential is also zero.

At (8,10) the value of the total differential is

$$df(8,10) = 2ydx + 2(x-y)dy$$
$$= 20dx + 2(8-10)dy .$$
$$= 20dx - 4dy$$

For the small changes ($dx = 0.1$ and $dy = 0.1$) that we chose, the total differential at (8,10) is

$$df(8,10) = 20dx - 4dy$$
$$= 20 \cdot 0.1 - 4 \cdot 0.1 .$$
$$= 1.6$$

Now, let's look at a real example where we use the exact differential to help understand physical phenomena. Let's grow crystals.

The equation[19] that describes the concentration along the growth axis of one element (or compound, depending on the materials) within another in a certain class of alloys is

$$C(x) = C_0 \left\{ 1 - (1-k)e^{-\frac{kR}{D}x} \right\},$$

where

$C(x)$ is the concentration of the component of interest as a function of the length, $x$, of the growing crystal

$C_0$ is the starting concentration of the component of interest, in the melt, before growth

$k$ is the ratio of the component of interest in the solid to that in the liquid at the growth surface

$R$ is the crystal growth rate. (To first estimate, it's the speed that the crystal is being pulled through the furnace from the melting region to the cooler part where the crystal forms upon solidification.)

$D$ is the liquid diffusion coefficient. It is a measure of mixing at the atomic level in the liquid.

---

[19] See "The Redistribution of Solute Atoms During the Solidification of Metals," W. A. Tiller, K. A. Jackson, J. W. Rutter, and B. Chalmers, *Acta Metallurgica*, vol. 1. page 428-437, 1953, and "Interface Shapes During Vertical Bridgman Growth of PbSnTe Crystals," Y. Huang, W. A. Debnam, and A. L. Fripp, *Journal of Crystal Growth*, vol. 104, pages 315-326, 1990, for more on this subject.

Typically you'll mix your starting chemicals, melt them, and then slowly pull your container from the hot (melt) zone to the cooler (growth) zone of the crystal growth furnace. Let's assume all does not go well. Your pulling apparatus has a hiccup, so you have a small change in growth rate. That is you get a $dR$. You forgot to set the timer so you aren't exactly sure where growth started, and you also have a $dx$. To make matters worse, your boss comes in and accidentally bumps the furnace causing a momentary change in the atomic mixing in the liquid which you model as a change in the diffusion coefficient—that is, you look at the bump as a $dD$.

Assume we can quantify these small changes ($dR$, $dx$, and $dD$), and you don't want your concentration to vary from the ideal by more than 5%. Has your growth been ruined? Let's see if

$$\frac{dC(x)}{C(x)} > 0.05 \ .$$

Let's first look at

$$dC(x, R, D) = \frac{\partial C}{\partial x} dx + \frac{\partial C}{\partial R} dR + \frac{\partial C}{\partial D} dD \ ,$$

Don't be confused because we now make the concentration a function of $R$ and $D$. $C$ has always been a function of $R$ and $D$; it's just that before we were holding $R$ and $D$ constant.

where

$$\frac{\partial C}{\partial x} = -C_0 k^2 \frac{R}{D} e^{-\frac{kr}{D}x} \ ,$$

$$\frac{\partial C}{\partial R} = -C_0 k^2 \frac{x}{D} e^{-\frac{kr}{D}x} \ ,$$

and

$$\frac{\partial C}{\partial D} = C_0 k^2 \frac{Rx}{D^2} e^{-\frac{kr}{D}x} \ .$$

Hence, the possible error is calculated as

$$\frac{dC(x)}{C(x)} = \frac{-C_0 k^2 \{\frac{R}{D} dx + \frac{x}{D} dR - \frac{Rx}{D^2} dD\} e^{-\frac{kR}{D}x}}{C_0 \{1 - (1-k)e^{-\frac{kR}{D}x}\}}$$

$$= \frac{-k^2 \{\frac{R}{D} dx + \frac{x}{D} dR - \frac{Rx}{D^2} dD\} e^{-\frac{kR}{D}x}}{\{1 - (1-k)e^{-\frac{kR}{D}x}\}}$$

We first note that the starting composition, $C_0$, drops out of the equation. We also note that we have $e^{-x}$ in the numerator and $1 - e^{-x}$ (with other factors) in the denominator. Since $e^{-x}$ gets very small as $x$ increases, we can, at large $x$, approximate the composition error equation as

$$\frac{dC(x)}{C(x)} = k^2 \{ \frac{R}{D} dx + \frac{x}{D} dR + \frac{Rx}{D^2} dD \} e^{-\frac{kR}{D}x} \, .$$

So, is the crystal ruined or not? You immediately see that it depends on how much has grown. The more that has already grown, that is: large x, the safer you are from those perturbations. If you wish to put in actual numbers, look up the referenced papers, and go for it.

Let's get really esoteric for another example and play with quantum mechanics. Quantum mechanics helps to describe what happens at the atomic, molecular, and lower levels. Civil engineers don't use it in designing bridges and petroleum engineers won't need it to drill through the earth's crust, but electrical engineers need to understand the inner workings of their integrated circuits, and chemists often work at the atomic level to make their new compounds.

Quantum mechanics is needed where energy is not continuous but only changes in finite steps or quanta. An early example of the use of quantum mechanics is when Niels Bohr (1912) looked at the energy of the electromagnetic radiation emitted by excited hydrogen atoms. He noted that regardless of the excitation induced onto the hydrogen, only certain energy levels were emitted from the hydrogen. He scratched his head, did lots of calculations, and came to the conclusion that the hydrogen's single electron circled the nucleus in only a limited number of fixed orbits. He postulated that the observed emissions occurred when electrons fell from higher (excited) orbits to lower (less energetic) orbits. The emitted energy was the difference in energy of the two orbits.

Bohr's model worked very well for the simple hydrogen atom, and it proved the usefulness of quantum theory. The theory, however, was not readily extended to atoms with more than one electron. A bit more than a decade later, the Austrian physicist, *Erwin Schrödinger* produced the famous equation that carries his name. His equation applied to a free electron in a potential well is

$$\left( \frac{\eta^2}{2m} \nabla^2 - V \right) \Psi(x,t) = -i\eta \frac{\partial}{\partial t} \Psi(x,t)$$

where

$\eta$ is the modified Planck's constant (It's a constant that Planck used to make one of his equations work. Don't worry about it.)

$m$ is the electron mass,

and

V is the height of the energy barrier.

The function we are trying to determine, $\Psi(x,t)$, is the wave[20] function. The wave function is a representation of the probability of finding the electron at a given position, $x$, at a given time, $t$.

The wave function has different solutions depending on where we are looking for our electron. If we are looking for the electron in the well, which we call region 2 in

Figure 8-6, where the barrier potential is zero, that is, where $V = 0$, the solutions are

$$\Psi_2(x,t) = Be^{ikx} + Ce^{-ikx}$$

If we are looking for the electron past the barrier[21] to the left (region 1), the solution is

$$\Psi_1(x,t) = Ae^{\alpha x}.$$

If we are looking in region 3, the solution is

$$\Psi_3(x,t) = De^{-\alpha x}$$

where $k$ is related to the square root of the energy, $E$, of the electron, and $\alpha$ is related to $\sqrt{V - E}$.

We didn't ignore the time dependent part. The time dependent parts canceled out.

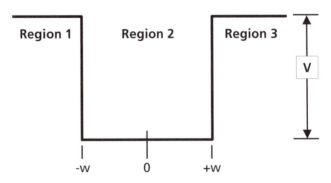

**Figure 8-6. An electron in a potential well.**

---

[20] Yes, we are treating electrons as if they are either particles or waves. That's OK, as experiments have shown that crystals can diffract electrons as if they are waves.

[21] What do we mean by 'Looking for the electron.' Remember, the wave function leads to the probability of finding the electron at some location. If there is no way that the electron will be at some location, then the value of the wave function for that position will be zero.

We expect the wave function to be continuous and smooth at the potential well walls which, in our figure, are at $x = +w$, and $x = -w$. (The width of the potential well is $2w$.) That is, we require

$$\Psi_1(-w) = \Psi_2(-w),$$

and

$$\Psi_2(w) = \Psi_3(w)$$

to satisfy the continuity requirement.

We also require

$$\frac{\partial \Psi_1(x)}{\partial x} = \frac{\partial \Psi_2(x)}{\partial x} \quad \text{when evaluated at } -w,$$

and

$$\frac{\partial \Psi_2(x)}{\partial x} = \frac{\partial \Psi_3(x)}{\partial x} \quad \text{when evaluated at } +w,$$

to satisfy the condition that the functions are smooth at the well boundaries.

Putting the equations into these requirements, we have

$$Ae^{-\alpha w} = Be^{-ikww} + Ce^{ikw}$$

and

$$Be^{ikww} + Ce^{-ikw} = De^{-\alpha w}$$

for the continuous part, and

$$-A\alpha e^{-\alpha w} = -ikBe^{-ikw} + ikCe^{ikw}$$

and

$$ikBe^{ikw} - ikCe^{-ikw} = -D\alpha e^{-\alpha w}$$

satisfy the condition that the first derivatives match at the boundaries.

Combining these equations, we get

$$(\alpha - ik)Be^{-ikw} + (\alpha + ik)Ce^{ikw} = 0$$

and

$$(\alpha + ik)Be^{ikw} + (\alpha - ik)Ce^{-ikw} = 0.$$

Using some more algebra we get

$$(\alpha^2 + k^2)B^2 = (\alpha^2 + k^2)C^2.$$

Assuming that $\left(\alpha^2 + k^2\right) \neq 0$ means that

$B^2 = C^2$, which is the same as saying that

$B = \pm C$.

Hence, inside the potential well the solutions to the wave equation are of the sort

$(e^{ikx} + e^{-ikx}) = 2\cos(kx)$

or

$(e^{ikx} - e^{-ikx}) = 2i\sin(kx)$.

Do not forget that the solution to the wave equation inside the well must match the solutions outside the well when the wave functions meet at the well boundaries. If we look at the boundary at $x = +w$, we have

$2B\cos(kw) = De^{-\alpha w}$

and

$2iB\sin(kw) = De^{-\alpha w}$

as possible solutions, with a similar set at $x = -w$. But we still have undetermined constants, and, more important, we don't know the possible values of k and $\alpha$ which tell us the allowed (remember: this is *quantum* mechanics) values of the energy of the electron.

Since we know that $B = \pm C$, we can manipulate our earlier equations

$\left(\alpha - ik\right)Be^{-ikw} + \left(\alpha + ik\right)Ce^{ikw} = 0$

and

$\left(\alpha + ik\right)Be^{ikw} + \left(\alpha - ik\right)Ce^{-ikw} = 0$

and obtain

$\dfrac{\alpha - ik}{\alpha + ik} = \pm e^{2ik\alpha}$

When we expand the imaginary exponential into trig functions and multiply the left-hand side by $\dfrac{\alpha + ik}{\alpha + ik}$ we get

$\dfrac{\alpha^2 - -2ik - k^2}{\alpha^2 + k^2} = \pm\left(\cos(2kw) + i\sin(2kw)\right)$.

Equating the real and imaginary parts we get

$\dfrac{\alpha^2 - k^2}{\alpha^2 + k^2} = \pm\cos(2kw)$

and

$$\frac{-2\alpha k}{\alpha^2 + k^2} = \pm \sin(2kw) \, .$$

Dividing the second equation by the first, we now have

$$\tan(2kw) = \frac{-2\alpha k}{\alpha^2 - k^2} \, .$$

Recall that $k$ is related by the square root to the energy, $E$, of the electron and also that $\alpha$ is related to the square root of the difference between the depth of the well, $V$, and the energy of the electron. Hence we see that only certain values of energy are allowed, and these energies are determined by the width, $w$, of the well and its depth, $V$.

## The Navier-Stokes Equations

Physics courses stress the fundamental law of conservation of momentum in a closed system. Momentum is defined as mass times velocity of that mass. As a simple example, note the high speed of a cue ball as it slams into the twelve racked balls at the start of a billiards game. After the cue ball hits the mass of balls, it might stop. The others will take off but at a slower speed of, on average, one twelfth the velocity of the cue ball before collision. That's conservation of momentum for the closed system on the thirteen pool balls. The only way to change the momentum of the system of balls is to apply an external force. That force was you with the cue stick.

Conservation of momentum of rigid systems such as pool balls is fairly straightforward (though the calculation with 17 balls will be difficult). Conservation of fluid systems is not so easily characterized. Here we turn to two greats of nineteenth-century mathematics and physics: Navier and Stokes.

Although the *Navier-Stokes* equations[22] were first postulated to describe the motion of nonturbulent Newtonian fluids (fluids where the shear stresses are a linear function of the fluid strain rate) in the nineteenth century, they have continued to be a subject that captures the interest of scientists. According to the database Web of Science (Science Citation Index), there is an average of 15–20 published papers per week that uses these equations of motion.

The *Navier-Stokes* equation, in the $x$-direction, for three-dimensional, incompressible flow can be written as

$$\frac{\partial u}{\partial t} + u\frac{\partial u}{\partial x} + v\frac{\partial u}{\partial y} + w\frac{\partial u}{\partial z} = -\frac{1}{\rho}\frac{\partial P}{\partial x} + \upsilon(\frac{\partial^2 u}{\partial x^2} + \frac{\partial^2 u}{\partial y^3} + \frac{\partial^2 u}{\partial z^2}) \, .$$

---

[22] Yes, it boils down to one equation, so our English major friends might want us to use the singular. However, this (these) equation(s) evolved over time; hence, there were different forms of The Equation at different times.

where

$u(x,y,z)$ is the velocity of the fluid in the *x*-direction,

$P$ is the pressure on the fluid,

$\rho(x,y,z)$ is the density of the fluid,

$\upsilon$ is the viscosity[23] of the fluid and *t* is time.

Note the that the term $\rho\dfrac{\partial u}{\partial t}$ is a force term. There are identical equations for the *y* and *z* directions.

Engineers who deal with atmospheric turbulence may note that the body forces[24] are not included. These are typically neglected in most engineering flows. While the equation includes a lot of terms, of particular interest are $\dfrac{\partial u}{\partial t}$, which is the change in velocity over time (the unsteady flow term); the $\dfrac{\partial u}{\partial x}$ which refers to the change in the x or downstream component of velocity in the *x* direction; and the $(\dfrac{\partial^2 u}{\partial x^2}+\dfrac{\partial^2 u}{\partial y^2}+\dfrac{\partial^2 u}{\partial z^2})$, which accounts for the momentum of the fluid flow.

If we assume that the flow is inviscid (i.e., zero viscosity, sometimes referred to as a perfect fluid), we are essentially asserting that there is no friction that we need to account for within the flow and that the only forces of note that are acting on the fluid are from the gravitational field. With this assumption, the daunting $\upsilon(\dfrac{\partial^2 u}{\partial x^2}+\dfrac{\partial^2 u}{\partial y^3}+\dfrac{\partial^2 u}{\partial z^2})$ term drops out of the equation.

The *Navier-Stokes* equation then reduces to $\dfrac{\partial u}{\partial t}+u\dfrac{\partial u}{\partial x}+v\dfrac{\partial u}{\partial y}+w\dfrac{\partial u}{\partial z}=-\dfrac{1}{\rho}\dfrac{\partial P}{\partial x}$, (often referred to as the Euler equation). If we also assume that the motion of the fluid is steady—that is, any $\dfrac{\partial}{\partial t}$ term is zero—the equation further simplifies to

$$u\frac{\partial u}{\partial x}+v\frac{\partial u}{\partial y}+w\frac{\partial u}{\partial z}+\frac{1}{\rho}\frac{\partial P}{\partial x}=0.$$

This form is referred to as the *Navier-Stokes* equation for inviscid flow. If we integrate over space which removes vector derivatives (see the chapter on vector calculus), and we break the pressure term into its components, we can derive the Bernoulli equation as follows:

---

[23] Consider viscosity as a "resistance to stirring" of the fluid. A heavy molasses has higher viscosity than water.

$$P + \frac{1}{2}\rho u^2 = C$$

where $C$ is a constant.

This is the equation that allows planes to fly. The shape of the airplane's wing forces a higher velocity of the air over the top of the wing than the bottom. Hence, according to the equation, the pressure on top of the wing decreases, and the plane rises. It's pretty neat that Bernoulli published his equation decades before Navier and Stokes finalized their equation and 165 years before Kitty Hawk.

Additional solved problems relating to engineering can be found on the accompanying CD-ROM.

---

[24] Forces that arise due to density changes.

CHAPTER **9**

# *Vector Calculus*

*A polar bear is a rectangular bear after a coordinate transformation.* — **Anonymous**

Vector calculus is yet another extremely useful engineering math tool. Let's begin our discussion of vectors with some vector manipulations you have been doing since kindergarten. As a little kid pulling a wagon, did you pull the handle straight up or did you pull in the direction that you wanted the wagon to go? When a friend wanted to help you, did you pull one direction and the friend pull another direction? No, you pulled in the direction that you wanted to go, and both you and your friend pulled in the same direction (assuming that your friends were more helpful than some of mine). You may not have put the words to it, but you were subconsciously maximizing the dot product of vectors to make that wagon move. The ***dot product*** is just a measure of the alignment (and magnitude) of your vectors: you want to pull your wagon in the direction that you want it to go. Sounds simple enough.

Sometime later, maybe in the second grade, you subconsciously maximized the cross product of vectors when you borrowed your parent's wrench and tightened the bolt connecting the handle to the wagon. A ***cross product*** is the mental calculation you did when you pulled that wrench at a right angle to its shaft.

Before we talk any more about dot and cross products—the stuff you've used since your early years—let's talk about simple vectors.

**Figure 9-1. Young child applying vector math to wagons.**

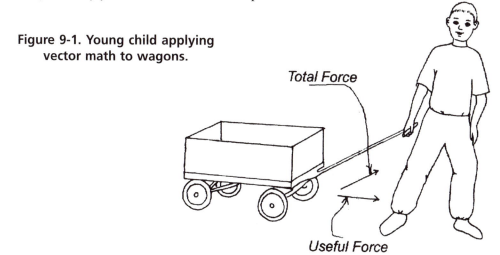

Total Force

Useful Force

A *vector* is a representation of a quantity and a direction[1]. You pulled the wagon with a force of a magnitude of perhaps two newtons (that's a quantity and there are about 5 newtons in a pound), and you pulled in the direction of your friend's house. So both the quantity and the direction of applying that quantity are important to accomplishing your goals. Let's put this in a sketch.

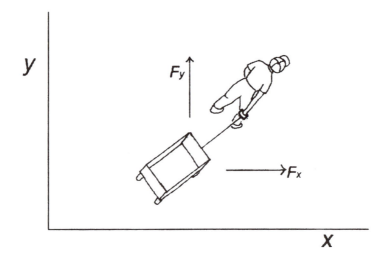

**Figure 9-2. Vector forces in pulling a wagon.**

In Figure 9-2 we show a vector representing your efforts to get the wagon from your house (which we put at the origin of the graph) to your friend's house (which is at some point in the distance). In the figure, we represent your effort with an arrow. Of course, we could have also presented your wagon-pulling efforts as

$$\vec{F} = \vec{F}_x + \vec{F}_y$$

where $\vec{F}_x$ is the force in the $x$ direction and $\vec{F}_y$ is the force in the $y$ direction[2]. The arrow[3] on top of $\vec{F}$ tells us that $\vec{F}$ is a vector.

---

[1]  If it's not a vector, it's a *scalar*. A scalar is a quantity without direction.
[2]  The use of the vector arrow and the directional subscript are redundant, and there is no need to use the arrow with the absolute magnitude signs. We'll soon stop this waste of our natural resources.
[3]  Some books will use a bold font ($\mathbf{F} = \vec{F}$) without an arrow to designate a vector.

You might look at Figure 9-2 and see that $\vec{F}$ points 45° off the *x*-axis. That's no surprise if your friend's house is somewhere in the direction of $x = y$, and you're taking the shortest route to get there[4]. Remember what you learned in trig and calculate that

$$\left|\vec{F_x}\right| = \left|\vec{F_y}\right| = \sqrt{2}\left|\vec{F}\right|$$

where $\left|\vec{F}\right|$ is the magnitude, or quantity, of $\vec{F}$ without regard to direction. And, of course, $\vec{F_x}$ and $\vec{F_y}$ are in the *x* and *y* directions, respectively.

Let's plot $\vec{F_x}$ and $\vec{F_y}$ on the same graph as $\vec{F}$ (Figure 9-3a). We see that if we place these two vectors with the tail of either one at the head of the other, we wind up with $\vec{F}$. Also note that vector addition is commutative. That is, given vectors $\vec{A}$ and $\vec{B}$, then

$$\vec{A} + \vec{B} = \vec{B} + \vec{A}.$$

Let's draw another graph with three vectors on it (also Figure 9-3a). Note that if we take these three vectors $\vec{A}$, $\vec{B}$, and $\vec{C}$ and add them in any order, we get to the same resultant vector. Hence, vector addition is associative. Or in symbols:

$$(\vec{A} + \vec{B}) + \vec{C} = \vec{A} + (\vec{B} + \vec{C}).$$

There are no surprises in those two laws, but be careful, some other vector operations do not follow the old rules of algebra. We'll get to them later.

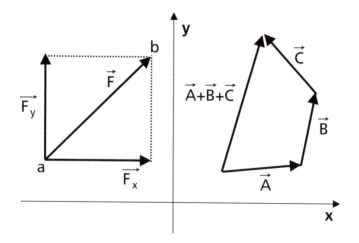

**Figure 9-3a. Vector Addition.**

---

[4]  And yes, we are mixing units in our coordinate system. One moment the coordinate system represents force, and in another moment it represents a map. Hang loose; it'll work out.

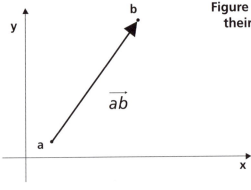

**Figure 9-3b. Vectors are defined by their magnitude and direction.**

We've talked about vectors that start at the origin of the coordinate system, and we've moved those vectors about to add them to other vectors. We can also describe a vector that connects any two points on a graph, such as points *a* and *b* as in Figure 9-3b. If we connect *a* and *b* in the direction from *a* to *b*, we write this vector as $\overrightarrow{ab}$, and likewise, if we went from *b* to *a*, the vector would be written as $\overrightarrow{ba}$. It should be obvious from the figure that

$$\overrightarrow{ab} = -\overrightarrow{ba}.$$

If you were pulling your little wagon with a force $\vec{F}$ and you got a spurt of energy and started pulling it with twice as much force in the same direction, your new force would be $2\vec{F}$. Thus, you can multiply a vector by a scalar. A **scalar** is a nonvector quantity that has magnitude but no direction.

We'll now introduce the unit vector. The **unit vector** is a vector with the length of one unit, in whatever units we're using. A unit vector can have any designated direction. We can make a unit vector out of any vector by dividing that vector by its magnitude. That is, a unit vector in the direction of $\vec{F}$ can be calculated by

$$\vec{U} = \frac{\vec{F}}{\left|\vec{F}\right|}.$$

It is also useful to identify the unit vectors that describe our coordinate system, but first let's expand our coordinate system. We've been working with the *x-y* system which describes a plane. Let's add a *z*-axis perpendicular to that plane so we can describe all space[5]. We'll add the positive *z*-axis in the direction that your thumb points if you take your right hand, place your open palm against the positive *x*-axis and sweep your hand toward the *y*-axis. The mechanics among you will recognize that the positive *z*-axis is the direction that a standard screw will advance if turned in the direction that you were sweeping your hand. Such a coordinate system is said to obey the **right-hand rule**. Remember this right-hand rule; you'll also see it in physics courses.

---

[5] And yes, we are ignoring Einstein's theories of relativity, black holes, and such. That's a topic for another book.

Now that we have our three-dimensional coordinate system, we'll introduce the unit vectors for that system: $\vec{i}, \vec{j},$ and $\vec{k}$, where $\vec{i}$ is the unit vector in the x-direction, $\vec{j}$ is the unit vector in the y-direction, and (we bet that you've guessed by now) $\vec{k}$ is the unit vector in the z-direction.

With these handy, orthogonal[6] unit vectors, we can describe any vector in terms of scalar magnitudes and $\vec{i}, \vec{j},$ and $\vec{k}$. For example, if you were a big kid when you pulled your wagon, your wagon handle might actually have an upward lift. Your pulling vector then might look like Figure 9-4. Then, you would write your vector as

$$\vec{F} = F_x\vec{i} + F_y\vec{j} + F_x\vec{k}$$

where $F_i, i = x, y, z$ are the scalar components of $\vec{F}$.

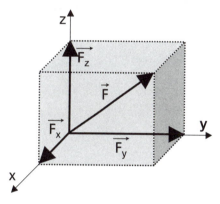

**Figure 9-4. Three-dimensional forces pulling on a wagon.**

## Solving Scalar Values

One method of solving these scalar values of the forces pulling the wagon is to first take the reading from the strain gauge connected to your arm (or just place a fisherman's spring scale on the wagon handle as you pull on the other end of the scale) to get $\left|\vec{F}\right|$. Measure the angles between the handle and the axes of your coordinate system to determine the directions of your force. The scalar components are then calculated as

$$F_x = \left|\vec{F}\right|\cos\alpha, \quad F_y = \left|\vec{F}\right|\cos\beta, \quad \text{and} \quad F_z = \left|\vec{F}\right|\cos\gamma$$

where $\alpha$, $\beta$, and $\gamma$ are the angles between $\vec{F}$ and the x, y, and z axes, respectively.

From the distance formula in the trig chapter, you know that

$$\left|\vec{F}\right| = \sqrt{F_x^2 + F_y^2 + F_z^2},$$

---

[6] Orthogonal is a word that vector calculus people like. It's the same thing as perpendicular or normal. The Cartesian coordinates are an example of an orthogonal axes system. We'll see other orthogonal systems later in this chapter.

Since $\vec{F}$ could also be a unit vector, we see that

$$\cos\alpha^2 + \cos\beta^2 + \cos\gamma^2 = 1.$$

From these relationships, we note that

$$\cos\alpha = \frac{F_x}{\sqrt{F_x^2 + F_y^2 + F_z^2}}$$

with similar equations for $\cos\beta$ and $\cos\gamma$.

We also note that the unit vector in the $\vec{F}$ direction is

$$\vec{U} = \vec{i}\cos\alpha + \vec{j}\cos\beta + \vec{k}\cos\gamma.$$

## The Dot Product of Vectors

The *dot product* of vectors $\vec{A}$ and $\vec{B}$ is defined as

$$\vec{A} \bullet \vec{B} = AB\cos\theta \qquad \text{Remember this one!} \quad \text{Eq. 9-1}$$

where

$A,\ B = \left|\vec{A}\right|,\ \left|\vec{B}\right|$, that is, the magnitude of the vectors, respectively,

and

$\theta$ is the angle between the two vectors.

The dot product is sometimes called the **scalar product** since it turns two vectors into a scalar quantity. Another lesser-used name for the dot product is the **inner product**.

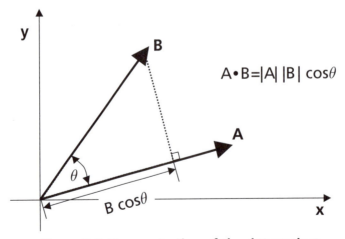

**Figure 9-5. Demonstration of the dot product.**

In words, the dot product is the projection of vector $\vec{A}$ onto vector $\vec{B}$ as shown in Figure 9-5. Please note that the projection of $\vec{A}$ onto $\vec{B}$ is the same as the projection of $\vec{B}$ onto $\vec{A}$, so

$$\vec{A}\cdot\vec{B}=\vec{B}\cdot\vec{A},$$

**Remember this one!** Eq. 9-2

In algebra-speak, we've just said that the dot product is commutative. Now, draw a few vectors and convince yourself that the dot product is also distributive. That is,

$$\vec{A}\cdot(\vec{B}+\vec{C})=\vec{A}\cdot\vec{B}+\vec{A}\cdot\vec{C}.$$

A few more obvious features of the dot product are:

1)   Should the angle be obtuse $(90° < \theta < 270°)$, the projection is taken as negative and the results acquire a minus sign.

2)   The dot products of the orthogonal unit vectors take on the following, obvious values[7]:

$$\vec{i}\cdot\vec{i}=\vec{j}\cdot\vec{j}=\vec{k}\cdot\vec{k}=1$$

and

$$\vec{i}\cdot\vec{j}=\vec{i}\cdot\vec{k}=\vec{j}\cdot\vec{k}=0.$$

3)   The dot product of a vector with itself produces the magnitude, squared, of that vector. That is,

$$\vec{A}\cdot\vec{A}=|A|^2.$$

With these relationships, note that if $\vec{A}$ and $\vec{B}$ are written as scalar coefficients of the orthogonal unit vectors, then

$$\vec{A}\cdot\vec{B}=(A_x\vec{i}+A_y\vec{j}+A_z\vec{k})\cdot(B_x\vec{i}+B_y\vec{j}+B_z\vec{k})$$
$$=A_xB_x+A_yB_y+A_zB_z,$$

and the angle between $\vec{A}$ and $\vec{B}$ is the inverse angle of

$$\cos\theta=\frac{A_xB_x+A_yB_y+A_zB_z}{\sqrt{A_x^2+A_y^2+A_z^2}\sqrt{B_x^2+B_y^2+B_z^2}}.$$

Taking the directional cosines of $\vec{A}$ and $\vec{B}$ as $\cos\alpha_A,\cos\beta_A,\cos\gamma_A$ and $\cos\alpha_B,\cos\beta_B,\cos\gamma_B$ verify for yourself that

$$\cos\theta=\cos\alpha_A\cos\alpha_B+\cos\beta_A\cos\beta_B+\cos\gamma\cos\gamma_B.$$

---

[7]  This might be a good time to introduce the so-called ***Kronecker delta***. The dot products among the orthogonal unit vectors could be written as $\vec{U_a}\cdot\vec{U_b}=\delta_{a,b}$, where $\delta_{a,b}$ is the *Kronecker delta* and is equal to 1 if $a=b$, and is otherwise equal to zero. The *Kronecker delta* is used in applications other than the dot products of unit vectors.

Would you like to rotate your coordinate system and see what your vector $\vec{F}$ looks like in that new system? Let's assume you are pulling your wagon from your home to your friend's house. Our existing coordinate system had your force vector (you pulling the wagon) in the direction $x = y$ and, since you are taller than the wagon, at an angle[8] $\gamma$ from the vertical z-axis. Now we are going to use a new coordinate system defined by three orthogonal unit vectors $\vec{u_1}$, $\vec{u_2}$, and $\vec{u_3}$. Although you changed your coordinate system, you did nothing to your vector. The force on the handle of the wagon did not change because you redefined an imaginary coordinate system, but components of $\vec{F}$ are different in the two different coordinate systems.

You first expressed your vector as

$$\vec{F} = F_x \vec{i} + F_y \vec{j} + F_z \vec{k} .$$

Now you want to express it as

$$\vec{F} = C_1 \vec{u_1} + C_2 \vec{u_2} + C_3 \vec{u_3} .$$

It looks messy, but the dot product comes to the rescue.

Using the properties of the dot products of orthogonal unit vectors, we see that

$$C_1 = \vec{F} \bullet \vec{u_1}, \quad C_2 = \vec{F} \bullet \vec{u_2}, \quad \text{and} \quad C_3 = \vec{F} \bullet \vec{u_3} .$$

Hence, $\vec{F}$ in your new coordinate system is

$$\vec{F} = (\vec{F} \bullet \vec{u_1})\vec{u_1} + (\vec{F} \bullet \vec{u_2})\vec{u_2} + (\vec{F} \bullet \vec{u_3})\vec{u_3} .$$

For example, rotate the z-axis of your original coordinate system by an angle $\gamma$ to align that axis with your aching arm[9] and call that the $\vec{u_1}$ direction. In that simple case, all of the force vector is in the $\vec{u_1}$ direction, and $\vec{F} = F\vec{u_1}$ .

## The Cross Product of Vectors

When you pull the wagon, you want your force vector aligned as much as possible with the vector representing the direction you want to go. However, when you try to tighten the bolts on the wagon, you want the wrench to be at right angles to your arm. In other words, when pulling the wagon, the only useful force is that projected onto a vector representing the desired direction of motion. When using the wrench, however, the only useful force is perpendicular to the wrench.

---

[8] Of course, that angle might be 90° or greater if you were a very little kid. Even so, the equations still hold. If you were a very big kid, and if the wagon was quite low to the ground, the angle from the z-axis would have been very small, and you could have carried the wagon. We won't bother with those cases.

[9] The old x-y plane would have to tilt by the same angle. The x and y axes, however, could rotate in that plane by any amount desired just as long as they remain orthogonal to each other.

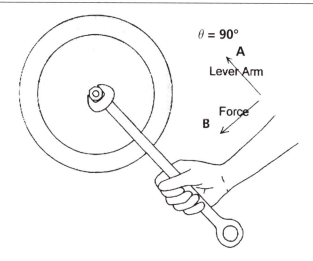

**Figure 9-6. Vectors demonstrating the cross product.**

Look at the two vectors $\vec{A}$ and $\vec{B}$ separated by an angle $\theta$ in Figure 9-6. The projection of $\vec{A}$ and $\vec{B}$ is the dot product. Similarly, the amount of $\vec{A}$ that is perpendicular to $\vec{B}$ is $A \sin \theta$. Since the length of the wrench is also important, the magnitude of the torque[10] is $AB \sin \theta$. This is the force that is tightening your bolt. And we're not through yet. We are working on an equation that describes the turn of a screw (or the turning of a nut on a bolt). Direction, in the turning of a screw, is very important.

With all that verbiage behind us, we'll define the cross product[11] of two vectors. The cross product of $\vec{A}$ and $\vec{B}$ is

$$\vec{A} \times \vec{B} = AB \sin \theta \vec{u}_c ,$$   **Remember this one!**   Eq. 9-3

where $\vec{u}_c$ is orthogonal to both $\vec{A}$ and $\vec{B}$ and points in the direction that a right-hand screw would advance if you started at $\vec{A}$ and swung towards $\vec{B}$ (see Figure 9-6).

Let's look at the orthogonal unit vectors[11] to clarify the cross product directions.

---

[10] Here is another physics term. Torque is the force on a lever arm. For those of you who've studied some physics, you'll note that torque and energy have the same units, which is force times distance. The difference is the relative directions (the directions on the unit vectors describing the force and the direction of movement) of the components. Energy is force times length when both are aligned in the same direction (such as your pulling the wagon to your friend's house). Torque is force times length where the two are perpendicular to each other (such as the bolt you were turning).

[10] Other monikers for cross product are vector product and outer product.

[11] $\vec{0}$ is the zero vector. Since the cross product produces a vector, we represent the result as a vector even if it has no magnitude. We'd also get a zero vector if we added $\vec{A}$ to $-\vec{A}$.

$$\vec{i} \times \vec{i} = \vec{j} \times \vec{j} = \vec{k} \times \vec{k} = \vec{0}$$

and

$$\vec{i} \times \vec{j} = -\vec{j} \times \vec{i} = \vec{k}, \quad \vec{j} \times \vec{k} = -\vec{k} \times \vec{j} = \vec{i}, \text{ and } \vec{k} \times \vec{i} = -\vec{i} \times \vec{k} = \vec{j}.$$

As stated, $\vec{i} \times \vec{j} = -\vec{j} \times \vec{i}$. This should be obvious from the definition of the cross product. For those of you who like big words, this means that the cross product is not commutative.

Don't try to memorize these relationships. Just visualize a *x-y-z*-coordinate system by holding your right hand, thumb up, in front of you. Your thumb is the *z*-axis, and your relaxed fingers point from the *x*-axis toward the *y*-axis. Once you draw, or at least visualize, this coordinate system, your curling hand will give you the answers. If you curl your hand from the direction of the *x*-axis (the $\vec{i}$ unit vector direction) toward the *y*-axis (the $\vec{j}$ unit vector direction), your thumb points in the direction of the positive *z*-axis which is in the direction of the $\vec{k}$ unit vector. If instead, you curl your fingers from the positive *x*-axis towards the negative *y*-axis your thumb points down in the negative $\vec{k}$ direction. Remember this relationship.

"That's fine," you retort, "but how does that handle the $\vec{i} \times \vec{i}$ relationships?"

That's easy. If you use your fingers to try to curl the *x*-axis onto itself, you make a fist. Fists are not appropriate in math, so it's a zero.

Now, let's calculate the cross product of two arbitrary vectors, $\vec{A}$ and $\vec{B}$. The calculation will look like a mess, but it's not.

$$\vec{A} \times \vec{B} = (A_x \vec{i} + A_y \vec{j} + A_z \vec{k}) \times (B_x \vec{i} + B_y \vec{j} + B_z \vec{k}).$$

Using the relationships shown above, we can drop terms derived from crossing like unit vectors and group the others with their resulting unit vectors to get

$$\vec{A} \times \vec{B} = (A_y B_z - A_z B_y)\vec{i} + (A_z B_x - A_x B_z)\vec{j} + (A_x B_y - A_y B_x)\vec{k}.$$

Exercise the fingers of your right hand and check this out. Then, look back at the chapter on matrices. Don't these terms look like the determinant,

$$\begin{vmatrix} \vec{i} & \vec{j} & \vec{k} \\ A_x & A_y & A_z \\ B_x & B_y & B_z \end{vmatrix} \quad ?$$

There's no magic here. It's just a useful coincidence that $\vec{A} \times \vec{B}$ equals the determinant formed by the orthogonal unit vectors and the components of each vector. Don't forget this useful relationship; we'll use it again.

Hence, $\vec{A} \times \vec{B} = \begin{vmatrix} \vec{i} & \vec{j} & \vec{k} \\ A_x & A_y & A_z \\ B_x & B_y & B_z \end{vmatrix}$ **Remember this one!** Eq. 9-4

Is the cross product useful for anything except tightening wagon bolts? Sure. Little David used it when he fought Goliath[13]. If David had curled his fingers to make a fist, he would have lost the fight. But David understood, at least in concept, the cross product. Effective use of a sling requires a high angular velocity and a long arm (in addition to good aim and a rock). The actual velocity of the rock leaving the sling is

$$\vec{V} = \vec{\omega} \times \vec{R}$$

where

$\omega$ is the angular velocity of the sling in revolutions per unit time

and

$\vec{R}$ is the length on the rotating arm and sling combined.

## Combining Vector Products

How many different ways can we combine *dot* and *cross* among three vectors? You might refer to the probability chapter and say we have two ways to choose the operation between the first and second vector. We also have two ways to choose the operation between the second and third vectors. This gives us four different combined vector products. You might think of the following operations:

1) $\vec{A} \cdot \vec{B} \cdot \vec{C}$

2) $\vec{A} \times \vec{B} \cdot \vec{C}$

3) $\vec{A} \cdot \vec{B} \times \vec{C}$

and

4) $\vec{A} \times \vec{B} \times \vec{C}$,

but you quickly see that 1) won't work without modification because once you take the dot product of any two of the vectors you have a scalar. Sure, you can multiply a vector by a scalar, but it isn't a dot product. A quick look at 2) and 3) shows similar potential problems, and 4) is confusing. Let's look at each one and see what works and what doesn't.

1) $\vec{A} \cdot \vec{B} \cdot \vec{C}$ is not defined, but $(\vec{A} \cdot \vec{B})\vec{C}$ and $\vec{A}(\vec{B} \cdot \vec{C})$ are nothing more than multiplying a vector by a scalar. Note that $(\vec{A} \cdot \vec{B})\vec{C}$ is not equal to $\vec{A}(\vec{B} \cdot \vec{C})$ except in a few special cases. The result, however, is a vector regardless of the order of operation. Among potential differences in these two results is that the final vector is in the $\vec{C}$ direction for the former operation and in the $\vec{A}$ direction in the latter operation.

[13] According to Manfred Korfmann in his October 1973 *Scientific American* article, "The Sling as a Weapon," the sling in the hands of a skillful person, such as a shepherd who used it frequently protecting his sheep from wolves, is a powerful weapon. Goliath didn't have a chance.

2) $\vec{A} \times \vec{B} \cdot \vec{C}$ is a perfectly good operation if we use our heads and cross before we dot. So, be safe: write $\vec{A} \times \vec{B} \cdot \vec{C}$ as $(\vec{A} \times \vec{B}) \cdot \vec{C}$, and keep everyone happy. The operation $(\vec{A} \times \vec{B}) \cdot \vec{C}$ is called the ***triple scalar product***, and the result is a scalar. Thus, since the dot produce is commutative,

$$(\vec{A} \times \vec{B}) \cdot \vec{C} = \vec{C} \cdot (\vec{A} \times \vec{B}).$$

3) $\vec{A} \cdot \vec{B} \times \vec{C}$ reflects the same discussion that we covered in 2) above. Would you believe that $\vec{A} \cdot (\vec{B} \times \vec{C})$ is equal to $\vec{C} \cdot (\vec{A} \times \vec{B})$ as well as to $\vec{B} \cdot (\vec{C} \times \vec{A})$? For the moment, accept this as true. Thus,

$$\vec{A} \cdot (\vec{C} \times \vec{B}) = \vec{C} \cdot (\vec{B} \times \vec{A}) = \vec{B} \cdot (\vec{A} \times \vec{C}).$$

These scalar results are the negative of those previously determined because we switched the order of crossing the two vectors.

To see these relationships, go back to the determinant representation on the cross product and dot it with the other vector. That is, if

$$\vec{B} \times \vec{C} = \begin{vmatrix} \vec{i} & \vec{j} & \vec{k} \\ B_x & B_y & B_z \\ C_x & C_y & C_z \end{vmatrix}$$

then

$$\vec{A} \cdot (\vec{B} \times \vec{C}) = \begin{vmatrix} A_x & A_y & A_z \\ B_x & B_y & B_z \\ C_x & C_y & C_z \end{vmatrix}.$$

If you doubt this result, work it out. It's a good exercise.

We have placed the triple scalar product into a determinant. A determinant is a determinant whether it comes from solving a group of connected equations, as we saw in the section on matrices, or from calculating dots and crosses[14]. The rules of determinants, therefore, hold in either case. Let's apply those rules to the triple scalar product.

---

[14] When we were working with two, three, or *n* unknowns back in the matrices chapter, we saw that we needed 2, 3, or *n* related, but not linearly dependent, equations to solve for those unknowns. We solved for those unknowns using a number of different techniques. One technique was the matrix method which required solving the determinate formed by the coefficients of the unknowns. Does it seem strange that we're now using determinants to solve the triple scalar product? We'll understand if you say yes.

Please do not think of the determinant as some magic operator we pull out of the air to facilitate mathematical operations. The determinant is nothing more than a handy way of writing a batch of equations. In multiple equations or triple scalar products, the determinant is a convenience, and the same rules apply to both operations. Think of the determinant as a handy symbol like the long division symbol you used back in grade school.

# Rules of Determinants

a) If one row is a linear combination of another row, the determinant is equal to zero. Two parallel vectors are linear combinations of each other. Hence, the triple scalar product is zero if any two vectors are parallel. Convince yourself of this by drawing a sketch.

b) If you interchange two rows, the sign on the determinant changes. By interchanging two rows you are changing the order of operation in the triple scalar product. Changing the order changes the sign as discussed in 3) above.

c) Interchange two rows, and you change the sign. Now interchange two more rows, and the sign changes again. It can only change back to its original sign. Thus, swapping rows twice leaves the sign on the value of the determinant unchanged. We hope you are now convinced we were telling the truth about all those different looking, triple scalar products of the same value.

4) The last multiple vector product that we'll work with involves two crosses. It's $\vec{A} \times \vec{B} \times \vec{C}$. A quick look at $\vec{A} \times \vec{B} \times \vec{C}$ should convince you that partitioning parentheses are necessary for this one, too. Why? If you cross $\vec{A}$ and $\vec{B}$ first and then cross that result with $\vec{C}$, the result will be perpendicular to $\vec{C}$. If, however, you cross $\vec{B}$ and $\vec{C}$ first, the result will be perpendicular to $\vec{A}$. It's possible they could be equal, but not probable.

4a) Let's cross $\vec{A}$ and $\vec{B}$ first and write it as $(\vec{A} \times \vec{B}) \times \vec{C}$. Of course, $\vec{A} \times \vec{B}$ is a vector. We'll call it $\vec{V}$, so crossing that vector with $\vec{C}$ also produces a vector. Since the result of all this crossing is still a vector, we'll call this operation the ***triple vector product***. (The discussion of the triple vector product is long and a bit tedious. Skip it if you wish, but if you stay the course, you'll really understand the cross product.)

$\vec{V} = \vec{A} \times \vec{B}$ is perpendicular to the plane defined by the vectors $\vec{A}$ and $\vec{B}$, so crossing $\vec{V}$ with another (nonparallel) vector will put the resulting vector (of this second crossing) back into the plane defined by $\vec{A}$ and $\vec{B}$. All of this circuitous argument says that $(\vec{A} \times \vec{B}) \times \vec{C}$ is a linear function of the vectors $\vec{A}$ and $\vec{B}$. That is,

$$(\vec{A} \times \vec{B}) \times \vec{C} = m\vec{A} + n\vec{B}$$

where $m$ and $n$ are constants. (We could have used unit vectors in the $\vec{A}$ and $\vec{B}$ directions. The magnitude of $m$ and $n$, for a general case, would have changed, but the value of the ratio of $m$ to $n$ would not change. Study this section again. It's subtle, but if you understand it, you understand the cross product.)

Do you want to calculate $m$ and $n$ in terms of $\vec{A}$, $\vec{B}$, and $\vec{C}$? We bet you've already figured out that the calculation involves the dot product since we must wind up with a scalar value for $m$ and $n$.

The first step we'll take in calculating *m* and *n* isn't necessary, but it will keep the paper work to a minimum. As you know, the vector sits out there in space[15], and we can rotate and translate the coordinate system (which is an arbitrary artifice of our imagination) anyway we want to. So let's rotate the coordinates such that thenew *x*-axis coincides with our vector $\vec{A}$. Since two noncoincident vectors define a ***plane***, we'll rotate the *y*-axis such that $\vec{B}$ is in the new *x-y* plane (see Figure 9-7).

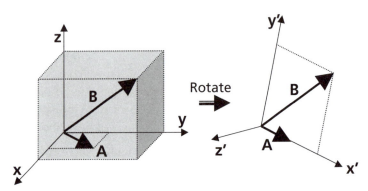

**Figure 9-7. Rotation of coordinate system to accommodate vectors.**

With those rotations of the coordinate system, vectors $\vec{A}$ and $\vec{B}$ can be written as

$$\vec{A} = A_x \vec{i}$$

and

$$\vec{B} = B_x \vec{i} + B_y \vec{j}$$

The first vector did not need a subscript on its magnitude since its action was only in the *x*-direction. We put it there for good bookkeeping.

Vector $\vec{C}$ can have any orientation with respect to the *x-y* plane, so we must write it as a general vector. That is

$$\vec{C} = C_x \vec{i} + C_y \vec{j} + C_z \vec{k} .$$

Using this coordinate system for our vectors, we have

$$\vec{A} \times \vec{B} = A_x B_y \vec{k} ,$$

which when crossed with $\vec{C}$ produces

$$(\vec{A} \times \vec{B}) \times \vec{C} = (A_x B_y \vec{k}) \times (C_x \vec{i} + C_y \vec{j} + C_z \vec{k})$$
$$= -(A_x B_y C_y) \vec{i} + (A_x B_y C_x) \vec{j}$$

---

[15] This space need not have length dimensions. We might be working in a space where the units are force, electric fields, or whatever you desire.

This vector is obviously in the $\vec{A} - \vec{B}$ plane as both $\vec{A}$ and $\vec{B}$ were written only in terms of $\vec{i}$ and $\vec{j}$. Note there is no contribution from the $C_x \vec{k}$ component of $\vec{C}$. Can you figure out why the $C_x \vec{k}$ component is ignored?

We can manipulate this equation into the form

$$(\vec{A} \times \vec{B}) \times \vec{C} = -(B_x C_x + B_y C_y)A_x \vec{i} + A_x C_x (B_x \vec{i} + B_y \vec{j}),$$

and then write it as

$$(\vec{A} \times \vec{B}) \times \vec{C} = -(\vec{B} \bullet \vec{C})\vec{A} + (\vec{A} \bullet \vec{C})\vec{B}.$$

We said that

$$(\vec{A} \times \vec{B}) \times \vec{C} = m\vec{A} + n\vec{B},$$

and now we see that

$$m = -\vec{B} \bullet \vec{C}$$

and

$$n = \vec{A} \bullet \vec{C}$$

We agree these steps are not obvious, but they are vector algebra. Go ahead and work backwards to convince yourself that it's right.

So, we have now solved for the scalar constants that we sought when we tried to simplify

$$(\vec{A} \times \vec{B}) \times \vec{C} = m\vec{A} + n\vec{B}.$$

Of course, we rotated the coordinate axes to facilitate the solution of $m$ and $n$. We can now rotate back to the original system if we wish. The equation will still be valid.

If we had crossed from the other direction, that is, had we wanted to simplify

$$\vec{C} \times (\vec{A} \times \vec{B}),$$

nothing would change but the signs of the $m$ and $n$ coefficients. If, however, you moved the parenthesis to produce

$$\vec{A} \times (\vec{B} \times \vec{C})$$

everything would change. The resulting vector would now reside in the $\vec{B} - \vec{C}$ plane. The solution for the scalar multipliers is still similar and is

$$\vec{A} \times (\vec{B} \times \vec{C}) = (\vec{A} \bullet \vec{C})\vec{B} - (\vec{A} \bullet \vec{B})\vec{C}.$$

Let's do one more exercise in vector algebra before moving to vector calculus. Let's play with four vectors and do a double-cross and a dot. Look at

$$(\vec{A} \times \vec{B}) \bullet (\vec{C} \times \vec{D}).$$

You recognize the necessity of crossing before dotting. If you calculated $\vec{B} \cdot \vec{C}$, you would have a scalar between two crosses and such operation is not defined. However, when writing this equation, we'll still put in the parentheses to avoid confusion. You also recognize that the result of these operations is a scalar since, by necessity, the last operation is a dot.

It may not be obvious, but $(\vec{A} \times \vec{B}) \cdot (\vec{C} \times \vec{D})$ is just a triple vector product inside of a triple scalar product. Follow these steps to see it happen.

Since $\vec{C} \times \vec{D}$ is a vector, we'll call that vector $\vec{V}$, and

$$(\vec{A} \times \vec{B}) \cdot (\vec{C} \times \vec{D}) = (\vec{A} \times \vec{B}) \cdot \vec{V}$$

which is a triple scalar product.

Remember how we switch the vectors about in the triple scalar product. So

$$(\vec{A} \times \vec{B}) \cdot (\vec{C} \times \vec{D}) = \vec{A} \cdot (\vec{B} \times \vec{V})$$

Expanding $\vec{V}$, we have

$$(\vec{A} \times \vec{B}) \cdot (\vec{C} \times \vec{D}) = \vec{A} \cdot [\vec{B} \times (\vec{C} \times \vec{D})]$$

which is the triple vector product inside of a triple scalar product.

We can simplify further, and get rid of all of the cross products. Go back to the section on the triple vector product, get the expanded form, and put it in the equation above to get

$$(\vec{A} \times \vec{B}) \cdot (\vec{C} \times \vec{D}) = \vec{A} \cdot [(\vec{B} \cdot \vec{D})\vec{C} - (\vec{B} \cdot \vec{C})\vec{D}].$$

When you run $\vec{A}$ through the right hand side of the equation, it becomes

$$(\vec{A} \times \vec{B}) \cdot (\vec{C} \times \vec{D}) = (\vec{A} \cdot \vec{C})(\vec{B} \cdot \vec{D}) - (\vec{A} \cdot \vec{D})(\vec{B} \cdot \vec{C}).$$

# Vector Calculus

As a rule, you differentiate vectors just as you did scalars, but, in some cases, you must be careful to follow the order of operations.

A vector that changes with respect to some parameter (such as a space vector that changes with respect to time) is differentiated as you'd expect. For example,

$$\frac{d\vec{V}(t)}{dt} = \frac{\vec{V}(t + \Delta t) - \vec{V}(t)}{\Delta t}, \text{ as } \Delta t \text{ gets very, very small.}$$

Likewise, if $\vec{V}(t)$ is modified by a scalar factor $s(t)$, then

$$\frac{d[s(t)\vec{V}(t)]}{dt} = s\frac{d\vec{V}}{dt} + \frac{ds}{dt}\vec{V}, \qquad \textbf{\textit{Remember this one!}} \quad \text{Eq. 9-5}$$

The $(t)$ is dropped on the right side of the equation to save ink (and tired typing fingers) in this and subsequent equations when the independent variable of the function is clearly understood.

If our vector is written as components along some coordinate axes such as

$$\vec{V}(t) = V_x(t)\vec{i} + V_y(t)\vec{j} + V_z(t)\vec{k}$$

where the orthogonal unit vectors define the coordinate system, the derivative of the function is

$$\frac{d\vec{V}}{dt} = \frac{V_x}{dt}\vec{i} + \frac{V_y}{dt}\vec{j} + \frac{V_z}{dt}\vec{k}.$$

If you want to differentiate equations with dots and crosses before you dot or cross, just go ahead and do it. Remember, never swap the order of the terms when the cross product is involved. (Shall we say, "Never cross your terms in a cross product.")

Examples:

$$\frac{d(\vec{A} \bullet \vec{B})}{dt} = \vec{A} \bullet \frac{d\vec{B}}{dt} + \frac{d\vec{A}}{dt} \bullet \vec{B},$$

$$\frac{d(\vec{A} \times \vec{B})}{dt} = \vec{A} \times \frac{d\vec{B}}{dt} + \frac{d\vec{A}}{dt} \times \vec{B},$$

and

$$\frac{d[\vec{A} \bullet (\vec{B} \times \vec{C})]}{dt} = \frac{d\vec{A}}{dt} \bullet (\vec{B} \times \vec{C}) + \vec{A} \bullet (\frac{d\vec{B}}{dt} \times \vec{C}) + \vec{A} \bullet (\vec{B} \times \frac{d\vec{C}}{dt}).$$

You might wonder what happens if a vector has a constant length but changes its orientation (direction) within the coordinate system. Let's look at that one by differentiating, in two different ways, the dot product of that vector with itself. That is, we'll first dot and then differentiate, and then we'll differentiate before we dot. Let $\vec{A}$ be a function of $t$, but only the direction of $\vec{A}$ changes. The magnitude of the vector is constant.

Dotting first we have

$$\frac{d(\vec{A} \bullet \vec{A})}{dt} = \frac{dA^2}{dt} = 0 \quad \text{(the derivative is equal to zero because } A^2 \text{ is the unchanging}$$

scalar magnitude of $\vec{A}$ )

and differentiating first we have

$$\frac{d(\vec{A} \bullet \vec{A})}{dt} = \vec{A} \bullet \frac{d\vec{A}}{dt} + \frac{d\vec{A}}{dt} \bullet \vec{A} = 2\vec{A} \bullet \frac{d\vec{A}}{dt}. \quad \text{(We can change the order of operations}$$

with the dot product.)

Combining the two different procedures we see that

$$\vec{A} \cdot \frac{d\vec{A}}{dt} = 0 \, .$$

So, either $\vec{A}$ has zero magnitude (that's not an interesting solution), it's not changing with respect to $t$, or $\vec{A}$ is orthogonal to its derivative. That is, $\vec{A}$ and $\frac{d\vec{A}}{dt}$ are at right angles to each other when only the direction changes with respect to the variable.

## The Gradient

Let's say you've been working too hard and you pull out the skis and the topo maps to plan a vacation. Each continuous line on the topo map shows an unchanging elevation above sea level. An adjacent line represents another elevation above sea level. The jump between each line represents an equal change in elevation. In vector calculus, these lines are called ***equipotential*** lines. Now, you don't care how far you are above sea level. You only want steep hills to conquer. What you want to know is the gradient of the mountain.

Each successive, equipotential line represents the same change in elevation. If these lines are close to each other, the terrain is steep, and good for skiing. You'll leave the areas where the lines are far apart to the novice skiers.

What does skiing have to do with vectors, you may be wondering, especially if you think that those equipotential lines look like scalars.

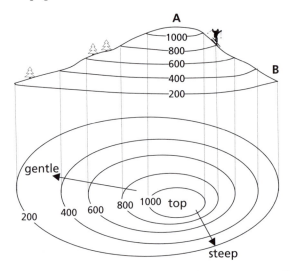

**Figure 9-8. Topo map developing the vector gradient.**

Yes, the equipotential lines are scalars, but you must choose a direction to go down that hill. You will go a distance in a chosen direction, and that is a vector. Look at the topo map sketched in Figure 9-8. Assume you want to go from the position and elevation marked A to the position marked B which, if you're skiing, is at a lower elevation. You will move a distance[16] of $d\vec{L}$ in going from A to B. If we let north be the $y$-axis and east the $x$-axis, then

$$d\vec{L} = dx\vec{i} + dy\,\vec{j}$$

where $\vec{i}$ and $\vec{j}$ are the orthogonal unit vectors[17].

The change in elevation as we move $dx$ and then $dy$ (or $dy$ and then $dx$) is

$$dh(x,y) = \frac{\partial h}{\partial x}dx + \frac{\partial h}{\partial y}dy.$$    **Remember this one!**    Eq. 9-6

We will now define the ***gradient***[18] of our lines as

$$\nabla h = \frac{\partial h}{\partial x}\vec{i} + \frac{\partial h}{\partial y}\vec{j},$$    **Remember this one!**    Eq. 9-7

which we call grad $h$. Our handy operator,

$$\nabla = \frac{\partial}{\partial x}\vec{i} + \frac{\partial}{\partial y}\vec{j},$$

is called the del operator.

---

[16] For those of you who want exactness: the dimensions on the north-south axis of a topo map are different from those on the east-west axis unless you are looking at a map of terrain on the equator. This is because at latitudes other than 0° the longitudinal lines are closer together than the lines of latitude. We'll try to not get tangled up in such things.

[17] Do you wonder what happened to the $z$-axis? We could think of the $z$-axis as the elevation we skiers are seeking, but that would complicate our thought process. The values of the equipotential lines determine the elevation on our map. If we could write a function for these equipotential lines (and we will write analytical functions for equipotential surfaces when we talk about more orderly things such as electromagnetic fields) the function for each line would look like $h(x, y) = C$, where $C$ is a constant. The function for the next line would be a different $C$, representing a different elevation.

Also, we may use all three axes if we have some physical phenomena that changes its magnitude due to its position in all three dimensions. The elevation of the hill only changes as a function of $x$ and $y$. The temperature may change not only with respect to where you are on a two-dimensional map but also may change as a function of the elevation ($z$-axis) above the ground. So, if we're talking about the temperature in three dimensions, the equi(temperature)potential surfaces would be $T(x, y, z) = C$. And, yes, we said "surfaces" not "lines" because we're looking at the temperature of a surface in our three-dimensional space and not the elevation of a line on our two dimensional map.

[18] This is the gradient of two-dimensional lines. The extension into three dimensions just adds a $\frac{\partial h}{\partial y}\vec{k}$ to the other terms.

We can now write the change in elevation as

$$dh(x, y) = \nabla h \bullet dl \ .$$

The gradient of our potential lines is the derivative of those potentials in each direction. Therefore, the gradient is a vector that points normal to the equipotential line and in the direction of the maximum change. Figure 9-8 will help demonstrate this concept.

# Divergence and Curl

So far, we have used vectors to describe the motion of wagons and the turning of wrenches. With a little imagination we can also use vectors to describe fluid motion. Let's go back to our childhood again and recall floating little boats in a stream of water. Neglecting the wind and our pushing them about, the boats move with the stream. The boats point in the direction of motion. This direction along with the boat's speed tells us the velocity of the stream. We have a vector representation of the fluid flow of our backyard creek.

Now let's note three things about our childhood experiences that will help in the understanding of *divergence* and *curl*.

1) In a gentle, unobstructed stream, two boats launched with a small separation between them will follow the same basic path. This observation describes *laminar flow*.

2) The amount of water in the stream does not change unless you are adding or subtracting water to it. When we quantify this observation in math-speak, we call it the *Divergence Theorem*. This means that we are assuming that the water does not compress under these conditions. If the stream widens or deepens, the velocity may slow down as compared to a narrower, shallower section, but the amount of water passing through the cross section of the stream in a given unit of time does not change. Steepening the gradient of the stream bed will also increase the flow velocity without changing the amount of water floating the boats. However, if rain falls, the sun is blazing and evaporating the water, you have an underground spring or a tributary, or if the cow is standing on the edge drinking water, the amount of flow will change.

3) If you build a partial dam or put some other object in your stream, one boat may spin around when passing that obstacle. The boat may even be caught in an eddy current downstream from your object and stay there going around and around[19]. The second boat may continue on downstream. We'll invoke the *curl* of a vector to help describe this action.

---

[19] Did you ever notice how hard it was to get one of your boats to repeat its exact path down a stream that was fast flowing and had many obstacles in it? As a child, you may have just enjoyed the challenge, but now you might ask why a physical experiment cannot be repeated when the overall stream flow does not seem to change. That is a really good question, and we'll skirt the answer later in the chapter on chaos.

Let's talk first about the divergence of a vector field. What we are doing is calculating the amount of fluid[20] added to or subtracted from the stream. We'll call something that adds fluid a ***source*** and whatever takes fluid away is called a ***sink***[21].

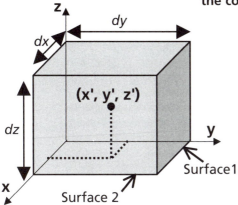

**Figure 9-9. Imaginary volume developing the concept of the divergence theory.**

Assume an imaginary rectangular volume, with sides $dx, dy, dz$, in the stream that encloses the point $(x', y', z')$ as in Figure 9-9. Let the fluid flow be represented by the vector field $\vec{F}(x, y, z)$. We can then approximate the change in each component of $\vec{F}$ as we move away from the center point $(x', y', z')$ to another nearby point as

$$F_x(x,y,z) = F_x(x',y',z') + \frac{\partial F_x}{\partial x}(x-x') + \frac{\partial F_x}{\partial y}(y-y') + \frac{\partial F_x}{\partial z}(z-z')$$

where we use the first term of the Taylor approximation[22] for $F_x$, and the partial derivatives are calculated at $(x',y',z')$. We ignore the higher-order terms of the expansion. We also have similar expressions for $F_y$ and $F_z$.

The point $(x',y',z')$ is at the center of the mini-volume, so the change in flow from the center to the surface #2 is

$$\int_2 F_x dS = dydz[F_x(x',y',z') + \frac{1}{2}\frac{\partial F_x}{\partial x}dx]$$ + those higher order terms that we ignore,

where $dydz$ is the area of surface #2.

---

[20] The math developed here is not restricted to fluid flow. It's also useful in electromagnetic theory.

[21] Some folks use the term negative source when referring to sink. To us that's like saying, "Put the dirty dishes in the negative source."

[22] The Taylor approximation was introduced in the chapter on calculus.

Note from the sketch that surface #2 is in the positive $x$ direction, so it represents the outflow of the $F_x$ component from $(x',y',z')$ through the surface in the $x$ direction.

Likewise, we have a flow into the volume element in the $x$-direction through surface #1 which is written as

$$-\int_1 F_x dS = -dydz[F_x(x',y',z') - \frac{1}{2}\frac{\partial F_x}{\partial x}dx] + \text{higher order terms that we'll also}$$

ignore.

Summing the flow in the $x$-direction, we have the change in flow through the volume, in the $x$-direction as

$$\frac{\partial F_x}{\partial x}dxdydz.$$

We can also run through the same argument for surfaces in the $y$ and $z$ directions, and determine the total change in flow through the volume as

$$\oint_S \vec{F} \cdot d\vec{S} = (\frac{\partial F_x}{\partial x} + \frac{\partial F_y}{\partial y} + \frac{\partial F_z}{\partial z})dxdydz,$$

where $\oint_S$ is the integral over the closed surface, and $\vec{F} \cdot d\vec{S}$ is the change in flow through the volume.

If we divide both sides by $dxdydz$, we'll have the net flow change per unit volume, and that is what we call the divergence of the vector $\vec{F}(x,y,z)$. This is written as

$$div\vec{F} = \frac{\partial F_x}{\partial x} + \frac{\partial F_y}{\partial y} + \frac{\partial F_z}{\partial z}.$$    **Remember this one!**  Eq. 9-8

If we recall the $\nabla$ operator which we used to determine gradients of scalar potentials, and dot it with our flow vector, $\vec{F}$, we get

$$\nabla \cdot \vec{F} = (\frac{\partial}{\partial x}\vec{i} + \frac{\partial}{\partial y}\vec{j} + \frac{\partial}{\partial z}\vec{k}) \cdot (F_x\vec{i} + F_y\vec{j} + F_z\vec{k})$$

which reduces to

$$\nabla \cdot \vec{F} = \frac{\partial F_x}{\partial x} + \frac{\partial F_y}{\partial y} + \frac{\partial F_z}{\partial z},$$

hence

$$div\vec{F} = \nabla \cdot \vec{F}.$$    **Remember this one!**  Eq. 9-9

# The Continuity Equation

A good example of the divergence theorem is the continuity equation. In plain language, the continuity equation says, "Stuff in equals stuff out plus or minus anything that happens to the stuff inside." Now, let's talk math.

Let's think back to our fluid stream. For the moment, we will not specify if the fluid is water, which we consider to be incompressible[23], or a gas which follows the perfect gas law. We are also assuming no sources or sinks in our volume.

If the mass density at a given point in three-dimensional space is $\rho(x,y,z)$, then the mass flow through a unit area perpendicular to the flow is $\rho\vec{V}$ where $\vec{V}$ is the velocity of flow at $(x,y,z)$. We can go through the same arguments we used to develop the divergence theorem for mass flow in and out. Now, however, we consider the possibility of something happening inside our test volume. We consider that the mass density might change. Hence,

$$\nabla \bullet \rho\vec{V} = -\frac{\partial \rho}{\partial t}.$$

**Remember this one!**  Eq. 9-10

This is the continuity equation[24]. Remember it. You'll see it again and again in engineering. But hang on, we're not yet through with it.

We'll throw a vector differentiation identity, without proof, at you. It is

$$\nabla \bullet \varphi\vec{U} = \varphi\nabla \bullet \vec{U} + \vec{U} \bullet \nabla\varphi$$

where $\varphi$ is a scalar function.

(If you desire, work it out. It's not that hard.)

Applying this identity to the continuity equation, we have

$$\nabla \bullet \rho\vec{V} = \rho\nabla \bullet \vec{V} + \vec{V} \bullet \nabla\rho = \frac{\partial \rho}{\partial t}$$

which can rearrange as

$$\rho\nabla \bullet \vec{V} = -\vec{V} \bullet \nabla\rho - \frac{\partial \rho}{\partial t}.$$

---

[23] Of course water is compressible. Its bulk modulus is $2.09 \times 10^9$ Pa. That's equivalent to saying that if an elephant (~5,000 kg) sits on a cube of water, with rigid side walls, with each surface area the size of your palm (~50 cm²), the column will compress ~0.5%. However, the small pressure changes we'll encounter in our flows are more like a small monkey sitting on the column. How does an elephant sit on that small cube of water? That's an engineering detail for you to solve later on.

[24] The minus sign is necessary because we are calculating the flow out of the volume element. An increase in density within the volume element will decrease the outward flow.

This can be messy working with a compressible fluid. If we have an incompressible fluid where the density doesn't change with position or time, then

$$\nabla \cdot \vec{V} = 0 .$$

This shows that the velocity vector of an incompressible fluid has zero divergence.

## Line Integrals

Before venturing into the curl, let's talk about line integrals. We discussed surface integrals when we introduced the divergence theorem. When working with a surface integral, we integrated a function over the bounding surface of a region. The surface integral is useful in many calculations ranging from how much sound comes from a surface to strength of attraction between two magnets. The surface integral is simply moving our standard integral into two dimensions. If the function is a vector, we can talk about *the number of flux lines crossing that surface*.

We will now integrate a function along a line. If $\vec{dl}$ is a minute vector along a path, then $\int_{1}^{2} \vec{F} \cdot \vec{dl}$, taken over the path $L$ from point #1 to point #2, is called the line integral. Sometimes we want to return to our starting point, that is, we want to go all around the block and return home. We call this closed loop line integral a **net circulation integral** and designate it as $\oint \vec{F} \cdot \vec{dl}$ .

Is the value of a line integral dependent on its path? Does it matter how we get from point 1 to point 2? Do we get different values if we choose different ways of going around the block? The answer to these questions is, "It depends."

Let's take an example from classic Greek mythology: the case of Sisyphus who pushes a rock up the hill only to have it come rolling back down again. We can examine this piece of mythological punishment by looking at its physics. In our frictionless example, the potential energy from gravity gets balanced with the kinetic energy of motion. The potential energy changes due to Sisyphus pushing a weight up a frictionless inclined plane. If we start with the weight sitting on the ground, we add potential energy to this system by raising it above its resting place. You may remember from your physics classes that potential energy is force times height. The force used in the potential energy is the weight of the stone, and the distance is that height (the slope of the hill doesn't matter except maybe to Sisyphus) above the initial starting point. The weight of the stone is just the counter force of a mass, $m$, times the acceleration due to gravity at the Earth's surface, represented as $\vec{g} = -g\vec{k}$ since the force is toward the center of the Earth, and we take that direction as $-\vec{k}$ . The energy changes only as we move the force parallel to its vector direction. We do not change the potential energy by sliding the weight along a flat surface level with the ground; we change the potential energy only by moving the weight in the $\vec{k}$ direction.

To put this in vector terminology, the energy change we are talking about is only potential energy, and that is

$$PE = m\vec{g} \bullet \vec{l}$$

Since the only component of $\vec{l}$ that will have a nonzero dot product with $\vec{g}$ is in the $\vec{k}$ direction, the change in potential energy is

$$PE = mgh$$

where

$h$ = the height above the ground (or up a hill) that the weight was raised.

Thus, neither the angle of the inclined plane nor the meandering path we took on the plane affects our answer. If we pushed that weight up the plane and allowed it to come back down—that is, if we had integrated over the closed loop—the final answer would have been zero.

Now, let's go to the real world. Friction exists even in Greek mythology. Since we have friction between the weight and the plane, we have additional energy requirements for Sisyphus pushing that weight on the plane even if it's not inclined upwards. We can make our point considering only the friction, so we'll ignore the other energy considerations even if that means we discount Sisyphus's sweat and aching bones.

The energy required to overcome the friction along the path is a *dissipative energy*, an energy primarily in the form of heat which is not readily put to useful work[25]. The dissipative energy, *DE*, due to friction is

$$DE = K(m\vec{g} \bullet \vec{n})l$$

where

$\vec{n}$ is the unit vector normal to the plane,

$K$ is the coefficient of friction,

and

$l$ is the length of the path.

---

[25] An interesting corollary to this discussion is found by looking at different automobile braking systems. The old way of stopping your car was by squeezing pads fixed to the frame of the car onto disks on the rotating wheels. If you don't think that a lot of useless heat was generated, touch one of those pads after a quick stop. The hybrid gasoline-electric automobiles use an electric generator connected to the wheels to stop. The velocity of the car is decreased by transferring kinetic energy (energy due to motion) to the generator that in turn recharges the car's batteries.

One of the key features to note in this equation is that *l* is *no longer a vector*! The equation for the potential energy featured a dot product with the displacement vector.

This resulted in the easy reduction of the line integral into the simple product of weight and height. The equation for the friction does not allow this simple reduction. As a result, we need to perform the line integral to determine the effects of friction.

Remembering your trig, you can calculate that the dissipative energy loss of going up our inclined plane is a function of the angle and length of the plane. We'll repeat the dissipative energy equation and then look at its terms.

$$DE = K(m\vec{g} \cdot \vec{n})l$$

$\vec{n} \cdot \vec{g} = g\cos\theta$, where $\theta$ is the angle of the plane above the flat earth,

and

$$l = \frac{h}{\sin\theta},$$

Hence,

$$DE = Kmgh\cot\theta.$$

The dissipative energy varies from zero for a straight-up lift to boundless for a nearly flat plane.

{Unfortunately, the need to describe a stone sliding down a perfect triangle is found about as often as someone suffering the eternal punishment of sliding a stone up and down an incline. Then again, perhaps we're being a little hasty. Some people pay a monthly membership fee at their health club for the privilege of sliding a heavy weight up and down an incline.

More often, we want to find the path between two points where the least amount of energy is dissipated. In other words, is it easier to walk over the mountain or around the mountain?}

## The Curl and Back to the Line Integral

Let's integrate an arbitrary function around a rectangular path in the *y-y* plane as shown in Figure 9-10. We'll start at the origin and first integrate along path #1, *dx*. Then we'll integrate along *dy* and so forth until we return to the origin. That is

$$\oint \vec{F} \cdot \vec{dl} = \int_1 F_x dx + \int_2 F_y dy - \int_3 F_x dx - \int_4 F_y dy$$

where the subscripts refer to the different paths.

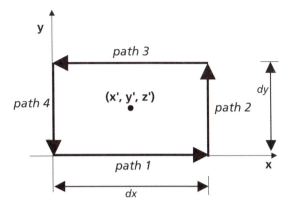

**Figure 9-10. The curl.**

If the point $(x',y',z')$ is at the center of our region described by our chosen path, we can expand the integral as

$$\oint \vec{F} \cdot d\vec{l} = F_x(x',y',z')dx - \frac{\partial F_x}{\partial y}\frac{dy}{2}dx + F_y(x',y',z')dy + \frac{\partial F_y}{\partial x}\frac{dx}{2}dy$$

$$-F_x(x',y',z')dx - \frac{\partial F_x}{\partial y}\frac{dy}{2}dx - F_y(x',y',z')dy + \frac{\partial F_y}{\partial x}\frac{dx}{2}dy .$$

Collecting like terms, we are left with

$$\oint \vec{F} \cdot d\vec{l} = (\frac{\partial F_y}{\partial x} - \frac{\partial F_x}{\partial y})dxdy .$$

If we divide through by *dxdy*, the area enclosed by our route, we have the value for the *circulation* per unit area of our route, which is $(\frac{\partial F_y}{\partial x} - \frac{\partial F_x}{\partial y})$. This is the curling in the *x-y* plane at the point $(x',y',z')$. The curling has direction. We define a positive value of the curl in the same manner as we defined the cross product: the right-hand rule. In this simplified example, we have the $\vec{k}$ component of the curl of the vector $\vec{F}$.

Using the above example, we can write out the three-dimensional curl of $\vec{F}$ as

$$curl\vec{F} = (\frac{\partial F_z}{\partial y} - \frac{\partial F_y}{\partial z})\vec{i} + (\frac{\partial F_x}{\partial z} - \frac{\partial F_z}{\partial x})\vec{j} + (\frac{\partial F_y}{\partial x} - \frac{\partial F_x}{\partial y})\vec{k} .$$

Is there an easier way to write and manipulate the curl? You bet. Look at the form of the curl as written, and think back to the cross product. Do you see the similarity? The curl is just the cross product of the del operator with your vector. That is

$$curl\vec{F} = \nabla \times \vec{F}$$

**Remember this one!** Eq. 9-11

and

$$\nabla \times \vec{F} = \begin{vmatrix} \vec{i} & \vec{j} & \vec{k} \\ \dfrac{\partial}{\partial x} & \dfrac{\partial}{\partial y} & \dfrac{\partial}{\partial z} \\ F_x & F_y & F_z \end{vmatrix}.$$

**Remember this one!** Eq. 9-12

Work it out to convince yourself it's true,

## Orthogonal Curvilinear Coordinates

Even though we used the normal $x,y,z$ coordinate system to define our vectors, the fundamental definitions of vectors and of operations of vectors with such things as curl, divergence, or gradient were not dependent on the specific coordinate system. Although the rectangular coordinate system may seem the most intuitive for many of our vectors, we may use other systems, and some physical phenomena will seem more intuitive in the other systems,

In our rectangular coordinate system, we start with a line which we usually call the $x$-axis. Perpendicular to (orthogonal to, or normal to), and connected to this line, we draw another line and call it the $y$-axis. We not only have two axes, but we have also defined a plane—the $x$-$y$ plane. Perpendicular to the $x$-$y$ plane, we draw another line—the $z$-axis. With these three lines, we can define all points[26] in space. We call this coordinate system the Cartesian coordinate system[27].

If we are not afraid of some messy calculations, we can define surfaces and volumes of cylinders and spheres in the Cartesian coordinate system, but there is an easier way to work with these shapes. We will define two new coordinate systems, the cylindrical coordinate system and the spherical coordinate system.

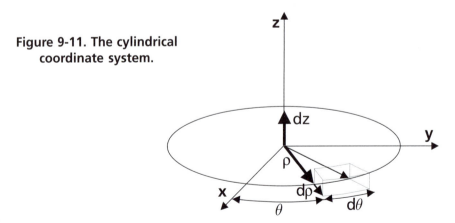

**Figure 9-11. The cylindrical coordinate system.**

---

[26] We'll leave relativistic physics out of this discussion.

[27] Other names for the Cartesian coordinate system are $x$-$y$, rectangular, and normal.

Consider a right cylinder such as a soup can sitting on the *x-y* plane as shown in Figure 9-11. Since all coordinate systems are just figments of the imagination, we can set our cylinder centered on the origin of the *x-y* axes. Our cylinder has a radius *R*. Recalling what we learned in trig, we define any point on the base of the cylinder as

$$x = \rho \cos \theta$$

and

$$y = \rho \sin \theta$$

where

$\rho$ is the scalar distance, in (or parallel to) the *x-y* plane, from the origin to the point under consideration (if we are talking about the label of the soup can, then $\rho = R$ )

and

$\theta$ is the angle, going counterclockwise, from the *x*-axis.

If we want to go above the base, we need a value for *z*.

From this short discussion, we hope it's obvious that any point in a Cartesian coordinate system can be described in terms of $\rho$, *$\theta$ and z*. The feat of going from *x,y,z* to $\rho,\theta,z$ is called ***transformation of coordinates***.

Example: The vector,

$$\vec{r} = 5\vec{i} + 5\vec{j} + 7\vec{k}$$

given in Cartesian coordinates converts to

$$\vec{r} = (5 * \sqrt{2} \cos 45)\vec{i} + (5 \cdot \sqrt{2} \sin 45)\vec{j} + 7\vec{k}$$

$$\rho = \sqrt{2} \cdot 5, \ \theta = 45°, \text{ and } z = 7$$

in cylindrical coordinates.

The most important thing to note is that unit vectors representing the directions of $\rho$, $\theta$ and *z* are all orthogonal to each other. The $\rho$-axis extends from whatever origin we choose out onto the *x-y* plane. The $\theta$ coordinate swings the $\rho$-coordinate, and is hence perpendicular to $\rho$, but it's still in (or parallel to) the *x-y* plane. The *z*-axis is normal to the *x-y* plane, and hence, normal to both $\rho$ and $\theta$.

If we want to define a cylindrical surface, we fix $\rho$ at *R*, the radius, we let $\pi$ vary from 0° to 360°, and we let *z* vary from 0 to *H*, the height of the cylinder. If we wanted to talk about the volume of the cylinder, the $\rho$ would vary from 0 to *R*. The bottom surface of the cylinder is defined as $\rho$ varying from 0 to R, $\theta$ varies from 0° to 360°, and *z* = 0. You figure the coordinates for the top of the cylinder.

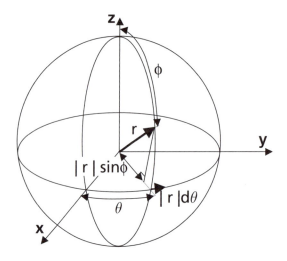

**Figure 9-12. Spherical coordinate system.**

If instead of using cylindrical coordinates where we drew our first new axis in the *x-y* plane before going vertical, we drew a vector directly to the point (*x,y,z*) as shown in Figure 9-12, we could define ***spherical coordinates***. Spherical coordinates are equivalent to latitude and longitude on the globe. In the figure, the vector swings down from the *z*-axis by $\phi$ degrees. (Note the difference in the $\rho$ vector in cylindrical coordinates and the *r* vector in spherical coordinates.) Hence the projection of *r* onto the *z*-axis is

$$z = r\cos\phi$$

where $|r| = \sqrt{x^2 + y^2 + z^2}$ .

The projection of *r* onto the *x-y* plane is $r\sin\phi$ which, similar to cylindrical coordinates, leads to

$$x = r\sin\phi\cos\theta$$

and

$$y = r\sin\phi\sin\theta .$$

If we hold *r* constant and let $\phi$ vary from 0° to 180°, and let $\theta$ vary from 0° to 360°, we have the surface of the earth. Convince yourself, as before, that *r*, $\phi$, *and* $\theta$ are orthogonal to each other.

Example:   The vector,

$$\vec{r} = 5\vec{i} + 5\vec{j} + 7\vec{k}$$

given in Cartesian coordinates converts to

$$|r| = 3 \cdot \sqrt{11} , \quad \phi = 45.3° , \text{ and } \theta = 45°$$

in spherical coordinates.

Before expressing some of our vector operations in cylindrical and spherical coordinates, let's do some integral calculus to see the usefulness of these systems.

If we want to determine the area of a rectangular surface in the *x-y* plane, we determine an elemental area, *dxdy*, and integrate over the limits of our surface. As an example, determine the area of a rectangle of *w* meters in *x*-direction and *l* meters in the *y*-direction.

Just for the fun of it, let's put the lower left corner of our rectangle on the origin of the Cartesian coordinate system, and we have

$$A = \int_0^l \int_0^w dx dy \,,$$

which we readily integrate to determine that $A = l*w$ m$^2$.

That was too easy. If we have some property, $f(x,y)$, of the rectangle that varies with position (such as a property that increased in value as we moved from the center of the rectangle) then the use of the double integral would have been worthwhile.

Now, let's play similar games with cylindrical coordinates. An elemental area on the *x-y* plane is now $d\rho$ in one direction, and orthogonal to that is $\rho d\theta$. Hence, the elemental area is $\rho d\rho d\theta$. Let's skip area calculations and go to volume. We need *dz* to complete the volume element of $\rho d\rho d\theta dz$. The limits of integration are $0 \le \rho \le R$, $0 \le \theta \le 2\pi$, and $0 \le z \le H$. The volume of a cylinder is

$$V = \int_0^H \int_0^{2\pi} \int_0^R \rho d\rho d\theta dz$$

**Remember this one!** Eq. 9-13

$$= \pi R^2 H$$

which you already knew. Similar to the rectangular coordinates, we could have had a function that represented some physical changes within the cylinder which we would integrate over the cylindrical coordinates.

Moving on to spherical coordinates, we have the volume element composed on the orthogonal components $r \sin\phi d\theta$, $rd\phi$, and *dr* which lead to the volume element of $r^2 \sin\phi d\theta d\phi dr$. When integrated, the volume of a sphere is

$$V = \int_0^R \int_0^\pi \int_0^{2\pi} r^2 \sin\phi d\theta d\phi dr$$

**Remember this one!** Eq. 9-14

$$= \frac{4}{3}\pi R^3$$

As an exercise, re-work the volume integral of a sphere and determine the area of the surface of a sphere which is $4\pi R^2$. (Hint: the length of $r$ does not change when you are only interested in the surface.)

Let's look at some of the vector operations in cylindrical and spherical coordinates. We'll start with the del operator

$$\nabla = \frac{\partial}{\partial x}\vec{i} + \frac{\partial}{\partial y}\vec{j} + \frac{\partial}{\partial z}\vec{k},$$

and follow the same reasoning we used in developing the elemental volumes. The del becomes

$$\nabla = \frac{\partial}{\partial \rho}\vec{a}_\rho + \frac{1}{\rho}\frac{\partial}{\partial \theta}\vec{a}_\theta + \frac{\partial}{\partial z}\vec{a}_z$$

for cylindrical coordinates, and

$$\nabla = \frac{\partial}{\partial r}\vec{a}_r + \frac{1}{r}\frac{\partial}{\partial \phi}\vec{a}_\phi + \frac{1}{r\sin\phi}\frac{\partial}{\partial \theta}\vec{a}_\theta$$

for spherical coordinates

where we use $\vec{a}_i$ to designate the unit direction vector in the $i$th direction.

If we apply the del operator to a scalar function, $U(\rho,\theta,z)$ or $U(\rho,\phi,\theta)$ we are determining the gradient of that function and the application is direct. That is

$$\nabla U = \frac{\partial U}{\partial \rho}\vec{a}_\rho + \frac{1}{\rho}\frac{\partial U}{\partial \theta}\vec{a}_\theta + \frac{\partial U}{\partial z}\vec{a}_z$$

for cylindrical coordinates, and

$$\nabla U = \frac{\partial U}{\partial r}\vec{a}_r + \frac{1}{r}\frac{\partial U}{\partial \phi}\vec{a}_\phi + \frac{1}{r\sin\phi}\frac{\partial U}{\partial \theta}\vec{a}_\theta$$

for spherical coordinates.

If we operate on a vector with the del operator, we are calculating the divergence of that vector. Remember, the divergence is a measure of the net outflow from a unit volume. We'll proceed to calculate the divergence in cylindrical coordinates in the same manner as in rectangular coordinates. We'll look at the flow through the various surfaces of an elemental volume.

Let's first consider the flow through the elemental surface normal to the $\rho$ direction. The flow through face 1 is

$$Flux_{\rho 1} = F_\rho \rho d\theta dz + \frac{1}{2}\frac{\partial F_\rho}{\partial \rho}\rho d\rho d\theta dz,$$

where we ignore higher-order terms as we did in rectangular coordinates.

The flow through the opposite surface is

$$Flux_{\rho 2} = -F_\rho \rho d\theta dz + \frac{1}{2}\frac{\partial F_\rho}{\partial \rho}\rho d\rho d\theta dz .$$

Hence, the net flow in the $\rho$ direction is

$$Net\ Flux_\rho = \frac{\partial F_\rho \rho}{\partial \rho}d\rho d\theta dz .$$

The net flow through the other two sets of surfaces are determined similarly. The net flux per unit volume emanating from the volume is defined as the divergence. Hence

$$\nabla \bullet \vec{F} = \frac{1}{\rho}\frac{\partial}{\partial \rho}(\rho F_\rho) + \frac{1}{\rho}\frac{\partial F_\theta}{\partial \theta} + \frac{\partial F_z}{\partial Z}$$

in cylindrical coordinates.

The divergence in spherical coordinates gets rather messy. We'll state it here and let you fill in the details.

$$\nabla \bullet \vec{F} = \frac{1}{r^2}\frac{\partial}{\partial r}(r^2 F_r) + \frac{1}{r\sin\theta}\frac{\partial}{\partial \theta}(\sin\theta F_\theta) + \frac{1}{r\sin\theta}\frac{\partial F_\phi}{\partial \phi} .$$

You can also determine the curl of a vector in cylindrical and spherical coordinates. The determination of these vector functions follows the determination in rectangular coordinates though the details are quite tedious. To save you the boredom of chasing all of the geometry, we'll just state the results.

Curl in cylindrical coordinates is

$$\nabla \times \vec{F} = (\frac{1}{\rho}\frac{\partial F_z}{\partial \theta} - \frac{\partial F_\theta}{\partial z})\vec{a_r} + (\frac{\partial F_r}{\partial z} - \frac{\partial F_z}{\partial r})\vec{a_\theta} + (\frac{1}{r}\frac{\partial (rF_\theta)}{\partial r} - \frac{1}{r}\frac{\partial F_r}{\partial \theta})\vec{a_z} .$$

The curl in spherical coordinates is

$$\nabla \times \vec{F} = (\frac{\partial (F_\phi \sin\theta)}{\partial \theta} - \frac{\partial F_\theta}{\partial \phi})\frac{\vec{a_r}}{r\sin\theta}$$

$$+ (\frac{1}{\sin\theta}\frac{\partial F_r}{\partial \phi} - \frac{\partial (rF_\phi)}{\partial r})\frac{\vec{a_\theta}}{r} + (\frac{\partial (rF_\theta)}{\partial r} - \frac{\partial F_r}{\partial \theta})\frac{\vec{a_\phi}}{r} .$$

# In-depth Example: *Electromagnetic Theory*

Now we'll use vectors to look at some electromagnetic theory.

As you probably know, like electrical charges repel and unlike charges attract each other. Charles Coulomb[28] quantified the repulsion/attraction of charges with his equation

$$\vec{F} = \frac{1}{4\pi\varepsilon_0} \; \frac{q_1 \, q_2}{r^2} \vec{a}$$

where:

$\dfrac{1}{4\pi\varepsilon_0}$ is equal to $8.99 \times 10^9$ newton-meter$^2$/coulomb$^2$,

$q_1$ and $q_2$ are the magnitude of the charges (in coulombs) under test,

$r$ is the distance (in meters) between the two charges,

and

$\vec{a}$ is a unit vector representing the direction between the charges.

The units of our force vector are, of course, newtons. Just look at the messy $\dfrac{1}{4\pi\varepsilon_0}$ factor and you see that the coulombs and meters in that factor cancel with the coulombs and meters in the equation.

If both charges have the same sign (either positive or negative), $\vec{F}$ is positive and the charges repulse each other. If the charges are of different signs, the force is negative and the charges attract each other.

Now that we understand Coulomb's law, we can define a vector for the electric field which is

$$\vec{E} = \frac{\vec{F}}{q_0}$$

$$= \frac{1}{4\pi\varepsilon_0} \; \frac{q}{r^2} \vec{a}_r$$

where $q_0$ is an imaginary test charge equal to one coulomb,

and

$\vec{a}_r$ is the unit vector pointing from the test charge $q_0$ toward any direction of interest.

---

[28] Charles Augustin de Coulomb, 1736-1806, was a French physicist. The unit of electrical charge is named after him. A coulomb represents a charge of approximately $6.26 \times 10^{18}$ electrons.

The units[29] of the electrical field are newtons/coulomb.

If we consider a point[30] charge of magnitude $q$ ($q$ is assumed positive) suspended by a nonconducting magical sky-hook, that point charge creates a field about it that repels other positive charges equally in all directions. That is, the vectors representing the field go in all directions as shown in Figure 9-13. Note that we draw the vectors in the direction another positive charge would try to move due to the electrical field generated by the first electric field.

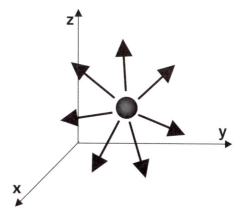

**Figure 9-13. The electric field
from a point charge.**

Now, let's bring up another electric charge of equal magnitude, but this time it's a negative charge. Set this charge a distance $d$ from the first one. Each charge has its own electric field. The vectors representing the field about our first charge point out from the charge, and the vectors point in toward the second charge. Out in space, the vectors add, but don't forget that they have opposite signs.

Let's determine the electric field along a line that bisects $d$, the separation of the two charges. Figure 9-14 shows the configuration. We put the charges at $\pm\dfrac{d}{2}$ on the $y$-axis, and we are asking about the field along the $x$-axis.

---

[29] The units of the electric field are also volts/meter. We are not playing games. Both sets of units are correct. It just means that a volt is equal to one newton times one meter/one coulomb.

[30] Of course, there is no such thing as a point charge. Electrons are small, but they occupy space. When physics types say *point whatever* they are looking at the effects of that point something at distances that are large compared to the dimensions of their thing.

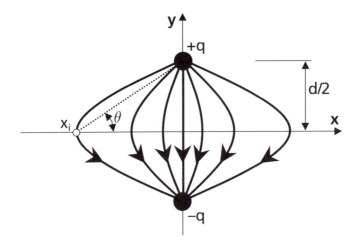

**Figure 9-14. The electric field from equal but opposite charges.**

If we ask about the field at the arbitrary point $x_i$ on the $x$-axis, we must calculate the contribution from both charges, and it is

$$\vec{E} = \frac{1}{4\pi\varepsilon_0} \left\{ \frac{q}{x_i^2 + \dfrac{d^2}{4}} \vec{a}_r - \frac{q}{x_i^2 + \dfrac{d^2}{4}} \vec{a}_r \right\}$$

$$= E_1\vec{a}_r - E_1\vec{a}_r$$

where $E_1$ is the magnitude of the field from either charge.

Let's look at these vectors in Figure 9-14, and break each into their $x$ and $y$ components as shown. At point $x_i$, the field from the positive charge is

$$-E_1(\cos\theta\,\vec{i} + \sin\theta\,\vec{j})$$

and the field from the negative charge is

$$E_1(\cos\theta\,\vec{i} - \sin\theta\,\vec{j})$$

where $\theta$ is defined in Figure 9-14.

Hence, you see by both the figure and math that the $x$-components of the electric field vectors cancel and $y$-components add. The electric field along the $x$-axis is

$$\vec{E} = -2E_1\cos\theta\,\vec{j}.$$

Now, let's do something about $\cos\theta$. Looking at Figure 9-14, we see that

$$\cos\theta = \frac{d/2}{\sqrt{x_i^2 + \left(d/2\right)^2}}$$

$$= \frac{d}{\sqrt{\left(4x_i^2\right) + d^2}} .$$

If desired, we can map this vector field in three dimensions about those two electric charges.

## Gauss's Law

You have just seen how to calculate a vector field from a scalar quantity in electro-magnetic theory. Do you wonder if the inverse is true? That is, if you know the vector field, can the charge that creates the field be determined? The answer is yes. The same Mr. Gauss that gave us a useful distribution in probability and statistics also gave us Gauss's Law for an electrical charge. It is

$$\varepsilon_0 \oint \vec{E} \bullet d\vec{A} = q .$$

where

$\varepsilon_0$ is the same constant that appears in the Coulomb force equation

$\vec{E}$ is the electrical field

$d\vec{A}$ is the elemental area of integration

$q$ is the enclosed charge

If you know the field around some area and can mathematically handle the geometry of that surface, you can calculate the enclosed charge. You'll see examples of this integration in advanced physics courses.

## The Lorentz Force

We refer to these examples as coming from electromagnetic theory, but all we've shown is the electro (charge) part of it. Are you ready for some magnetic vectors? Did you ever wonder how your computer or TV screen manages to produce[31] pictures? You may know that an electron beam dances about and lights up a phosphor on the back of the viewing screen, but do you know what plays the tune for the electron beam dance? The Lorentz[32] force equation quantifies the combined forces on a charge[33] due to an electric field and a magnetic field. His equation is

$$\vec{F} = q\vec{E} + q\vec{v}\times\vec{B} .$$

---

[31] We are talking about the tube type screen, not the flat plasma or liquid crystal screen.

[32] Named for Hendrik Lorentz (1853-1928), a Dutch physicist.

[33] The charge is just an electron in our computer or TV. That charge might be all sorts of things if we're talking about the earth's magnetic field deflecting solar particles from hitting us.

The $\vec{F} = q\vec{E}$ is just Coulomb's law. This part of the force equation[34] describes the initial motion of the electron as outlined in Figure 9-15. The $\vec{F} = q\vec{v} \times \vec{B}$ (where $\vec{v}$ is the velocity of the charge, and $\vec{B}$ is the magnetic field) part is new. It says that a charge speeding through a magnetic field[35] will move at right angles to the plane defined by the velocity vector and the magnetic field vector

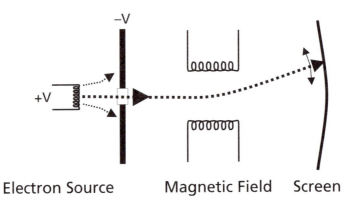

−V

+V

Electron Source       Magnetic Field     Screen

**Figure 9-15. A schematic of a display tube and the Lorentz force. The magnetic field, $\vec{B}$, is perpendicular to the plane of the paper.**

We made Figure 9-15 simple with the magnetic field at right angles to the electron's velocity vector. In operation, there will be another magnetic field at right angles to the one shown. Varying the strengths of the two magnetic fields will guide the beam across as well as up and down the screen.

## Force on a Current Loop

If you should have a loop of wire sitting in a magnetic field, you can make an electric motor by passing electrical current through the loop. Figure 9-16 shows the set up.

The equation that quantifies the force on the wire is just an extension of the Lorentz force equation and is

$$\vec{F} = i\vec{L} \times \vec{B}$$

---

[34] We show a voltage creating the electric field. That's OK. Remember that the units of the electric field are also volts/meter. Think of the voltage as equipotential lines. These are not elevation equipotential lines above sea level that will accelerate a skier going down the hill, but are equipotential lines that will accelerate an electric charge. The gradient of the equipotential lines is the electric field.

[35] You may already know that a magnetic field is a vector field. When you play with magnets you see direction and polarity (sign) properties.

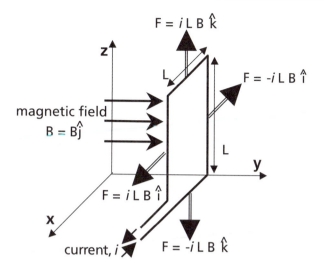

**Figure 9-16. A simple motor made by passing current through a loop of wire in a magnetic field.**

where

$i$ is the current,

$\vec{L}$ is the length of wire,

and

$\vec{B}$ is the magnetic field.

We made our loop in the shape of a rectangle in Figure 9-16, and the face of the loop is perpendicular to the magnetic field. That is, the loop is in the *x-z* plane only and $\vec{B} = B\vec{j}$. To make life even easier, let each segment of the loop be $L$ in length with each leg aligned with either the *x* or *z* axis.

We want to know the force on this loop. In each segment, the magnitude of the force is

$$|F| = |i\vec{L} \times \vec{B}|$$

$$= |iLB|$$

but magnitude alone doesn't tell the story. Force is a vector quantity. Hence the forces in the segments are

$$\vec{F}_{1,x} = iLB\vec{k},$$

$$\vec{F}_{2,z} = iLB\vec{i}, \text{ (don't confuse the current, } i, \text{ with the unit vector in the } x\text{-direction, } \vec{i} )$$

$$\vec{F}_{3,x} = -iLB\vec{k},$$

and

$$\vec{F}_{4,z} = -iLB\vec{k} \, ,$$

which obviously, sums to zero. Doesn't sound like a very efficient electric motor does it?

Now let's change the loop orientation in Figure 9-16 to that in Figure 9-17 where the loop has been rotated into the *x-y* plane by the means of mechanical connections midway up the vertical (*z*-axis) legs of the loop. We now want to calculate the torque on the current loop, but first we'll review the forces.

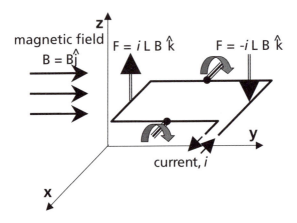

**Figure 9-17. Torque on a current loop where the plane of the loop is parallel to the magnetic field.**

$\vec{F}_{2,z}$ and $\vec{F}_{4,z}$ are now zero as their length vectors are parallel to the magnetic field, and $\vec{F}_{1,x}$ and $\vec{F}_{3,x}$ are unchanged other than they are offset from each other. They now do more than just push against each other. They produce a torque about the midpoint of the mechanical connections.

As you may recall, torque is that twisting action of a force acting at a distance. Torque about a given point is

$$\vec{T} = \vec{r} \times \vec{F}$$

where $\vec{r}$ is the length vector from our point of reference to the point where the force is applied.

The torque produced by segment one is

$$\vec{T}_1 = \tfrac{1}{2}\vec{L} \times \vec{F}$$
$$= \tfrac{1}{2}iL^2 B\vec{k}$$

and the torque on segment three is

$$\vec{T_3} = \tfrac{1}{2}\vec{L} \times \vec{F}$$
$$= \tfrac{1}{2} i L^2 B \vec{k} \,.$$

Hence, the total torque about the mechanical connections is

$$\vec{T} = \vec{T_1} + \vec{T_2}$$
$$= i L^2 B$$

Does this look like the start of an electric motor? Yes!

If the current loop were at an angle $\theta$ (as measured from the x-y plane to segment 2 of the loop), the torque arm would shorten to $l = L \cos\theta$, and the torque equation becomes

$$\vec{T} = i L^2 B \cos\theta$$

The action of torque is to rotate. The torque of the current loop in the magnetic field is trying to rotate our loop from the x-y plane ($\theta = 0°$) where the torque is twisting as hard as it can to the x-z plane ($\theta = 90°$) where the torque is zero.

Look at any real motor and you will see more than one current loop in it. Two loops at right angles to each other will insure a continuous torque at the mechanical connections. And more loops will linearly increase $L$.

Incidentally, if you apply an external rotating force (torque) to a current loop in a magnetic field, you'll produce an electrical current. That's a generator.

## Projectile Motion, Part 1.

Let's do some shooting. We know the velocity of our particle is $V_0$ as it leaves the launcher, and it travels over our flat range until gravity makes it hit the ground. We ignore wind resistance. Let's describe its flight and figure out how to maximize its distance.

Since we didn't specify the angle of the launch we can assume an angle, $\theta$, with respect to the ground and specify its initial trajectory as

$$\vec{V_0} = V_x\vec{i} + V_y\vec{j}$$
$$= V_0 \cos\theta\,\vec{i} + V_0 \sin\theta\,\vec{j}$$

Once in the air, the trajectory is only affected by gravity, $g$, so,

$$\vec{V}(t) = V_0 \cos\theta\vec{i} + (V_0 \sin\theta - gt)\vec{j}\,.$$

We see that the vertical component will decrease in magnitude until it reaches its peak height where it stops and starts to descend. The horizontal component of flight keeps chugging along unchanged until the $y$-component causes it to hit the ground. How high and how far it goes for a given $V_0$ depends on $\theta$. Let's look at how $\theta$ controls the action.

The distance traveled is velocity times time, so

$$X = V_0 \cos\theta \cdot T$$

and

$$Y = V_0 \sin\theta \cdot T - \int_0^T gt\,dt$$

where $T$ is the time of measurement.

Since you are near the surface of the Earth, you can assume that $g$ is constant (9.8 m/sec$^2$), so

$$Y = V_0 \sin\theta \cdot T - \frac{1}{2}gT^2 .$$

"OK," you respond, "We have equations for distance in $x$ and $y$, but how do we find the angle to gives us maximum range?"

The answer is to look at the values of $t$ (time) when $y$ is equal to zero. The position of the vertical component of flight is zero when the projectile is first fired, and it's zero again when it hits the ground. When it hits the ground, it has reached its maximum range (bounces don't count).

$y$ is zero, at the end of flight, when

$$V_0 \sin\theta \cdot T = \frac{1}{2}gT^2$$

or

$$T = \frac{2V_0 \sin\theta}{g} .$$

Of course, $y$ is zero if $\theta$ is zero, but we want a more interesting answer. We'll put this equation for $T$ into the equation for the range, $X$, and get

$$\begin{aligned}
X &= V_0 \cos\theta \cdot T \\
&= V_0 \cos\theta \cdot \frac{2V_0 \sin\theta}{g} . \\
&= \frac{2V_0^2}{g}\sin\theta\cos\theta
\end{aligned}$$

This is still messy, so go back to your trig identities and see that

$\sin\theta\cos\theta = \frac{1}{2}\sin(2\theta)$.

With simplification we now have

$X = \dfrac{V_0}{g}\sin(2\theta)$

which is maximum at $\theta = 45°$.

## Projectile Motion, Part 2.

Now let's launch a satellite. We know that the earth is curved. We know that gravity makes any object fall toward the center of the earth. Hence, our satellite must go fast enough tangentially to the earth's surface to remain at a constant distance from the earth even as gravity is pulling on it. The rocket science of this example is shown in Figure 9-18. We won't worry about such details as atmospheric resistance.

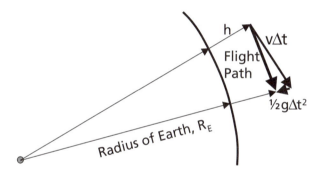

**Figure 9-18. Schematic of a satellite in free fall about the earth's surface.**

We shoot[36] the rocket up to whatever height we want, turn it in the direction we want it to orbit, and give it enough of a boost that it stays at the desired altitude above the earth even as earth's gravity is pulling it down. The rocket scientists call this *free fall* about the earth's surface. The rocket folks also calculate their trajectories using centrifugal force and other factors to determine their orbital mechanics. We can get a good idea of what's going on from projectile motion.

---

[36] Did you ever wonder why NASA uses a launch site in Florida? It's not just to be closer to Disney World. The closer they are to the equator, the more the rotation of the earth gives them a boost. If NASA had a launch site on the equator, they would get a free addition of speed of 1667 km/hr (assuming that they want to circle at the equator) over launching from Santa's home.

Figure 9-18 is expanded in Figure 9-19 to show a triangle composed of three vectors: one vector represents the earth's radius plus a small length addition to account for the satellite's altitude. Another vector represents the tangential distance traveled by the satellite in time, $\Delta t$. The third vector accounts for the radial distance that the satellite falls during $\Delta t$. To complete the triangle, the vector for the center of the earth to the satellite is repeated to connect to the fallen satellite.

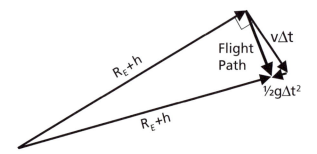

**Figure 9-19. Expansion of Figure 9-18.**

We look at the vectors and see that $\overrightarrow{R_E + h}$, $\overrightarrow{v\Delta t}$, *and* $\overrightarrow{R_E + h + \frac{1}{2}g\Delta t^2}$ approximate a right angle for small angles.

where

$\overrightarrow{R_E}$ is the vector from the center of the earth to the earth's surface (6370 km),

$\vec{h}$ is the normal vector from the earth's surface to the satellite,

$\overrightarrow{v\Delta t}$ is the tangential distance traveled by the satellite in time t,

and

$\frac{1}{2}g\overrightarrow{\Delta t^2}$ is the distance the satellite fell[37] in time $\Delta t$.

---

[37] *g*, of course, is the acceleration due to gravity, which is approximately 9.8 m/s² at the earth's surface, and varies as $\frac{1}{R^2}$. For low earth orbit, say 200 km, the difference in distances makes only a small change in the acceleration due to gravity. So you ask, "What's this about zero gravity in space? I've seen the astronauts floating about while in orbit."

No, your eyes didn't deceive you. But they were not floating as much as they were falling. They and the space shuttle are in free fall about the earth.

Hence,

$$(R_E + h)^2 + (v\Delta t)^2 \approx (R_E + h + \tfrac{1}{2}g\Delta t^2)^2$$
$$\approx (R_E + h)^2 + g(R_E + h)\Delta t^2$$

which reduces to

$$v = \sqrt{g(R_E + h)} \ .$$

If you put in 200 km for $h$ you'll get an orbital velocity of approximately 28,000 km/hr.

# Computer Mathematics

*The control of large numbers is possible, and like unto that of small numbers, if we sub-divide them.* —**Sun Tzu, 6th–5th century BC Chinese general, author of *The Art of War***

Mathematicians have spent centuries feverishly working to apply math to solve something useful. Unfortunately, that task has proven to be very difficult. If the physical system under investigation involves perfect shapes, then the mathematicians succeed[1] and the equations can be solved. Real situations rarely involve perfect shapes. Real life features strangely shaped structures experiencing unusual disturbances. Real-life problems are often very difficult to solve exactly with the mathematics expounded in this book.

However, don't despair. While we might not be able to solve the system exactly, we can often approximate the solution. The complicated shapes and complicated disturbances of real life can be subdivided into little pieces. These little pieces can be solved with the mathematics in this book. Keeping track of all of these little pieces is just laborious bookkeeping. Fortunately, computers are a perfect tool for this type of bookkeeping. The entire world of numerical mathematics, ranging from finite elements to differential equation solvers and from structural mechanics to population models, is simply a search for a way to approximate the complicated system with a bunch of little simple equations and the bookkeeping to keep track of the interaction between the bunches of simple equations.

This chapter will talk about how you can use your computer to solve math problems. We will not make you an expert on finite difference and finite element techniques, but hopefully you will gain a working knowledge. This chapter will talk about things that are easily modeled on the computer and things that are difficult to model on the computer. There are also things that are impossible to model on the computer, but we will leave those alone. Finally, this chapter should show you how you can use a spreadsheet to write simple, but useful simulations of real life.

## Computerized Data Analysis

One of the most straightforward reasons to use the computer is to help analyze data. This is what you do when you balance the checkbook. In equation form, the balancing of the checkbook can be expressed as:

---

[1] One problem in modeling physical systems is determining the acceptable trade-off between simplifying shapes and especially boundaries into something that is mathematically tractable and what is physically significant. An example of this quandary occurred when the mathematician was asked to quantitatively describe how a cow could jump over the moon. "An easy problem," was the reply, "First we assume a spherical cow..."

$$B = S + \sum_n deposits - \sum_i debits$$

where $B$ is the final balance, $S$ is the starting balance, and there are $n$ deposits and $i$ debits in the bank account. The summation sign, $\sum$, means that you are adding all of the terms together. The computer cranks through the transactions, performing the addition and subtraction for you. In effect, you are integrating your bank transactions, and the result of the integration is your balance. This is probably the most common type of computer math.

Along with using the computer to track your data, you can also use it to fit the data to a theoretical equation that is often a complicated mess that no sane person wants to work out with pencil and paper. An example that one of the authors suffered with is shown at the end of this chapter.

You can also use the computer to compute integrals and differentiate measured data. Let's suppose that you are drilling an oil well, and you are measuring the motion of your drill bit with an accelerometer. You can use the computer to integrate the acceleration data with respect to time and determine the velocity. You can integrate the acceleration data twice, with respect to time, and find the displacement. In other words, you can determine if you are about to drill through your oil field by numerically integrating the acceleration measured on your drill bit.

Computerized integration can be performed with different degrees of accuracy. The most accurate solution, the acceleration, $a$, is integrated with respect to time, $t$, to find the velocity, $v$.

$$v(t) = \int a(t)dt .$$

The velocity is just the integral of the acceleration. If the acceleration is constant, then the equation reduces to

$$v = v_o + at ,$$

where $v_o$ is the initial velocity.

## Finite Difference Technique

The simplest approach is to assume that everything keeps going until you make a new measurement. In evaluating the integral, we break the integral into small enough pieces that we can assume that the acceleration is constant. The computerized[2] form of the equation is then

$$v\langle t + \Delta t \rangle = v\langle t \rangle + a\langle t \rangle \Delta t ,$$

where $\Delta t$ is the amount of time between each measure of the acceleration.

---

[2] Some folks write $v\langle t + \Delta t \rangle$ as $v\langle t + 1 \rangle$, where the numeric simply tells the number of time steps the function has moved from the original $t$.

When going into the computer, time gets broken into chunks or discretized[3]. Instead of being continuous, time is measured when we measure the rest of the data. As a result, we use the bracketed $<t>$ to indicate that the value on those variables exists at the same instant[4] in time. Variables with a bracketed $<t + \Delta t>$ occur at an instant of time, $\Delta t$, later.

Similarly, the position, $p$, can be found by integrating the velocity or double integrating the acceleration,

$$p(t) = \int v(t)dt = \iint a(t)dtdt.$$

From our discussion of integrals, we know that if the acceleration is constant, then

$$p(t) = p_o + v_o t + \tfrac{1}{2} at^2$$

The integral can be discretized for use in the computer by phrasing (grouping) it as

$$p\langle t + \Delta t \rangle = p\langle t \rangle + v\langle t \rangle \Delta t.$$

Since the velocity is a function of acceleration, we can expand the equation as

$$p\langle t + \Delta t \rangle = p\langle t \rangle + \left( v\langle t \rangle + \tfrac{1}{2} a\langle t \rangle \Delta t \right) \Delta t,$$

or

$$p\langle t + \Delta t \rangle = p\langle t \rangle + v\langle t \rangle \Delta t + \tfrac{1}{2} a\langle t \rangle \Delta t^2$$

---

[3] Discretized vs. digitized: do not confuse these two words. The real world is analog (ignore quantum mechanics for this discussion) where things move and change continuously. We, however, do not measure physical phenomena continuously. We measure when and where we want to measure. If we are methodical, we measure at predetermined intervals of time, space, or whatever else may be our variable. These measurements are still determinations of analog data, but once measured, we may then digitize our data. In digitizing the data, we are putting numbers on the measurements. As an example, if you measure the air pressure in your car's tires, you put the gauge on the tire's valve and a little rod pokes out of the end of the gauge. The greater the pressure, the further the rod extends. You have made an analog measurement. If you measure the pressure once a week, your discretization period is one per week (some discretized measurements integrate and average the measurement over the entire period – that's hard to do with tire gauges). Now you read a number on your rod (it might extend to the line that reads $2 \times 10^5$ Pa) and write down that value. That is a digital measurement. The computer will want ones and zeros for its digital data.

[4] While we're on the subject of discretized time measurements, let's talk about proper choice of intervals. Essentially, the faster you expect the data to change, the more measurements you need to make. As an example, assume that you are measuring a sine wave that has a period of one second. If you make a measurement every second, you'll get the same value every time, and you won't know that you have a sine wave running about. Information theory teaches us that the absolute minimum time intervals that will give you useful data is half of the period of the highest frequency (highest frequency means shortest period of repetition) that you want to measure. For our 1-Hz sine wave, we must measure twice per second. Sketch it out and see if you agree with us.

Note that this form is identical to that found by evaluating the integral as we did earlier. What about this mysterious factor of ½ that magically appeared? Well, it is not entirely mysterious because we know from the integration that there should be a ½ before the $at^2$. This reflects that we need to use the average velocity over the time period. This is the type of thing that is easy to perform on a spreadsheet.

Is this the only way to integrate? Of course not. If this were the only way to integrate, then we wouldn't have computer science departments and expensive mathematical codes. You can rephrase the integration to try to predict how the acceleration might have behaved between the measured data. One way of doing this is by placing your knowledge of the system into the equations; then you can get a more accurate estimate of the position. Another way of predicting the acceleration uses extensive previous records. In our example of drilling an oil well, you are limited by how quickly you can send data to the surface. Thus, the more complicated integration techniques become very important.

You can also differentiate with the computer. Let's look back at the checkbook example. If you have the intermediate balances from your checkbook, then you can differentiate to see how your cash flow is going. If we have measured the displacement of our oilfield logging tool, then we can differentiate to determine how quickly it is moving. Differentiation is key for designing the cruise control on your car to keep the speed from wildly oscillating. We differentiate position measurements in order to determine the strain on a structure and to find whether failure is imminent.

How does differentiation work? All we have to do is return to how we defined differentiation. Remember, the derivative is the slope of the function, the tangent to the curve, the rate of change. In math parlance, the velocity (the derivative of position) is:

$$v = \frac{df(t)}{dt}$$
$$= \frac{f(t + \Delta t) - f(t)}{\Delta t}$$

as $\Delta t$ becomes very small.

The definition of the derivative can be directly translated into the computer notation. If you have measured data, then the derivative at time, $v<t>$, is

$$v\langle t \rangle = \frac{f\langle t \rangle - f\langle t - \Delta t \rangle}{\Delta t},$$

where $f<t>$ is the data measured at time $t$ and $f<t - \Delta t>$ is the data measured one time period earlier. If you measure the data often, then $\Delta t$ will be small and our new equation for the derivative will become an increasingly accurate measure of the derivative. As $\Delta t$ becomes smaller, you will also magnify problems associated with measurement noise. As a result, there are more complicated phrasings for the derivative that take data values over a longer period of time to more accurately interpolate the derivative.

The second derivative in time would give us the acceleration from measured displacement data. On a structure, the second derivative in space of the displacement of a structure would be the curvature. We will derive the expression for the acceleration because we'll be applying it during the finite elements section of this chapter. The acceleration, $a$, is the second derivative change of position with respect to time:

$$a = \frac{d^2 f(t)}{dt^2}$$

$$= \frac{dv(t)}{dt}$$

$$= \frac{v(t + \Delta t) - v(t)}{\Delta t}$$

as $\Delta t$ becomes very small.

Since the velocity is the first derivative of the displacement function, then the acceleration can also be expressed as the time rate of change in the velocity. In other words, the acceleration measures how quickly your velocity changes.

We have now reduced the second derivative into being the derivative of the first derivative. In other words, we have now reduced the problem into one that we have just solved. Substituting yields

$$a\langle t \rangle = \frac{v\langle t \rangle - v\langle t - \Delta t \rangle}{\Delta t}$$

where $v<t>$ is defined above and $v<t - \Delta t>$ is the same except that one is subtracted from the time increment, i.e.

$$v\langle t - \Delta t \rangle = \frac{f\langle t - \Delta t \rangle - f\langle t - 2\Delta t \rangle}{\Delta t} .$$

Substituting for $v<t>$ and $v<t - \Delta t>$ into the acceleration yields

$$a\langle t \rangle = \frac{\left( \left( f\langle t \rangle - f\langle t - \Delta t \rangle \right) \Big/ \Delta t \right) - \left( \left( f\langle t - \Delta t \rangle - f\langle t - 2\Delta t \rangle \right) \Big/ \Delta t \right)}{\Delta t}$$

$$= \frac{f\langle t \rangle - 2f\langle t - \Delta t \rangle + f\langle t - 2\Delta t \rangle}{\Delta t^2}$$

Just as with the first derivative, you can include more data points into the calculation of the acceleration in order to minimize the effects of signal noise.

Where will the computer get you in trouble? As the sage said, "To err is human, but to truly foul things up requires a computer." The computer can get you in trouble when you try to integrate or differentiate. The easiest route to disaster lies in our measurement error. If we are integrating data, then a slight offset in our measurement will add up to a large error. By offset in the measurement, we mean that we measure a small value when we should be measuring zero. In time, that constant small value will integrate into a big value. As a result, the integration of acceleration data is most often used over only very short periods of time or if a high-pass filter can be applied to remove the offset error. If we are differentiating, then this small offset will disappear.

Signal noise is the Achilles heel of differentiation. Signal noise appears as a variation on the true measurement. When we go to differentiate the data, then the small errors from the signal noise can be magnified as we shrink our sampling interval. Integration smoothes out the signal noise and, thus, signal noise tends not to be a problem when integrating data.

## Optimization Techniques

As we've mentioned before, the computer is great at rapidly performing boring calculations. This feature can be used to optimize the solution to a problem. What type of problems can you optimize? Just about anything. You can optimize distribution networks, optimize feedback controllers, find the roots of complicated equations, or determine the best way to cut cookies out of cookie dough. The key to successful optimization techniques is the definition of a cost function.

All optimization techniques use a cost function. The cost function yields a single value that you are trying to minimize. If you are running a business, then the cost function would calculate your profits. If you are designing a distribution network, then the cost function would calculate the total time it would take to move the products. Alternatively, the cost function for the distribution network could measure the total expense of the distribution. Equally valid would be to define the cost function as being a weighted combination of time and expense.

You have to be careful when you define your cost function. If you are not careful, then the computer will find the minimum of the cost function, but this minimum will not give you useful information. The computer is a stupid beast of burden. It will do *exactly* what you *tell* it to do, not what you *wish* it would do.

There are many ways to define the cost function. The most common form of a cost function features a value squared. This is sometimes called the 2-norm, but don't worry about the name. Let's say that we want to find the roots of the function $f(x,y,z)$. The cost function, $J$, would be

$$J = \left(f\left(x,y,z\right)\right)^2.$$

When the cost function *J* is 0, then we have a root of $f(x,y,z)$. Note that the cost function can been defined with multiple variables.

"Fine," you may say, "I have a cost function, but how do I actually find its minimum?"

There are as many answers to that question as there are mathematicians in the universe. Okay, maybe not that many, but there are many different solution techniques. We're going to talk about two of them: 1) the shotgun approach and 2) a gradient descent approach.

The simplest solution technique is called the shotgun approach, or more professionally a Monte Carlo simulation[5]. In a Monte Carlo simulation, the cost function is evaluated at random values of the variables. In other words, you try a bunch of values and hope that one of them is close to the optimum. This technique works well if you don't know much about the cost function, but this technique tends to require a large number of iterations. Expect to spend a long time number crunching if you have to run a Monte Carlo simulation.

The gradient descent optimization is an alternative to the Monte Carlo simulation. The gradient descent optimization starts when you pick a point in the solution space. Then you draw a line tangent to the curve at this point. For the next guess, you follow that tangent line until it reaches zero. That zero crossing[6] becomes your next guess. This process iterates until your next guess is very close to your previous guess. For most functions, the gradient descent is a quick optimization technique. You can define a function where the gradient descent will exhibit ghastly behavior, but generally you have to try rather hard.

In the gradient descent optimization, you are behaving as a drop of water that falls down the mathematical hill. The optimum is found when water stops falling downhill and collects in a puddle. Unfortunately, you don't know if your water has collected in a high mountain pond, in the ocean, or in the Dead Sea (408 meters below sea level). Where the water stops depends on where you initially placed it. In other words, gradient descent optimizations will find you a minimum, but it may only be a local minimum. As a result, different initial locations are often tried when using a gradient descent optimization. Alternatively, a Monte Carlo simulation might be performed in order to get an approximate value for the global minimum and then a gradient descent optimization is performed from the best solution from the Monte Carlo simulation.

---

[5] Monte Carlo simulations are named after the gambling tables of Monte Carlo, Monaco. With the Monte Carlo simulation, you are effectively rolling the mathematical dice and gambling that you will arrive at a satisfactory answer to your optimization.

[6] Note that if you had a few points along a linear curve and then followed this technique, you would obtain the root of the equation in your first try. We know that it's obvious, but it makes the point that the gradient technique has merit.

One of the variations, the genetic algorithm, puts some randomness into the gradient descent algorithm in order to avoid being trapped in local minimums. There are many, many different types of optimization techniques that are variations of the gradient descent algorithm.

## Finite Element Techniques

Let us start by saying that finite elements are not a separate type of mathematics. Like all of the other types of computer math, the finite element techniques are simply a method for expressing a complicated system as a series of simple equations.

The difficulty associated with the finite element techniques is in keeping track of all of the elements. There tend to be so many elements that special tricks are needed to be able to speed the solution. Additionally, as we will see, the form of the equations tends to allow special solution techniques.

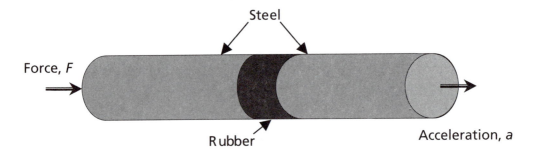

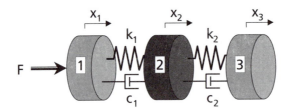

**Figure 10-1. Spring-mass-damper system.**

Let's start by looking at a vibration moving through a section of pipe that has a rubber isolation mount in it. We want to find the movement at one end of the pipe from a disturbance at the other end. This type of analysis would be very difficult to perform analytically. Using the computer greatly simplifies the analysis.

The first step is to separate the more complicated structure into simpler components. We can approximate the continuous rod as a series of small masses, springs, and dampers. The sum of the small masses should equal the total mass of the rod, and the combination of the small springs should equate to the stiffness of the rod.

The dampers or dashpots represent friction and internal dissipation of the material. The more small pieces we include, the higher the accuracy of the system.

Now we have a series of small problems that we can solve. The physics is simply

$$F = ma$$
$$= m\frac{d^2x}{dt^2}.$$

The next step is to find which forces act on which block. A free-body diagram is used to look at the forces on each component. In a free-body diagram, the forces acting on each element of the system are drawn and denoted. Making free-body diagrams is something that's taught in physics and engineering classes, and the results are shown in Figure 10-1.

This next section takes the free-body diagram and extracts a series of equations from it. The equations are all based upon the basic relationship, $F = ma$, or switching the notation,

$$ma = F$$

and on to the form you saw in the differential equations chapter.

$$m\frac{d^2x}{dt^2} = Kx + C\frac{dx}{dt} + F(t).$$

The forces, $F$, are the spring loads, $kx$, and the damping loads, $c\,dx/dt$, that are denoted in Figure 10-2. The bold notation is used to show that these variables are vectors and matrices. Substituting these forces into the equation yields

$$\begin{bmatrix} m_1a_1 \\ m_2a_2 \\ m_3a_3 \end{bmatrix} = \begin{bmatrix} -k_1 & k_1 & 0 \\ k_1 & -k_1-k_2 & k_2 \\ 0 & k_2 & -k_2 \end{bmatrix}\begin{bmatrix} x_1 \\ x_2 \\ x_3 \end{bmatrix} + \begin{bmatrix} -c_1 & c_1 & 0 \\ c_1 & -c_1-c_2 & c_2 \\ 0 & c_2 & -c_2 \end{bmatrix}\begin{bmatrix} \frac{dx_1}{dt} \\ \frac{dx_2}{dt} \\ \frac{dx_3}{dt} \end{bmatrix} + \begin{bmatrix} F(t) \\ 0 \\ 0 \end{bmatrix}$$

**Figure 10-2. Forces in the spring-mass-damper system.**

The equations are written in matrix format because matrices greatly simplify the notation. Additionally, many of the computer programs are vectorized, which means that everything runs much faster if it is phrased as a combination of vectors and matrices.

There are a couple of interesting things to note about the form of these matrices. The first is that the matrices tend to be symmetric; the transpose[7] of the matrix is the same as the original. Another thing to note is that most mass, stiffness, and damping

---

[7] You get the transpose of a matrix when you interchange its rows and columns.

matrices feature a series of diagonal arrays that are nonzero although the arrays contain many zeros. Much of the work of finite element modeling comes from exploiting the large number of zero elements in these matrices.

"Okay," you may say. "So I have the equations. How do I solve them?" Getting the equations is the physics and the engineering. While we could solve these equations analytically for this simple case, more complicated situations would not suggest themselves to an analytical solution.

We're going to use the definition of the acceleration from the computerized data analysis section to move our differential equation into being a finite difference equation. For a reminder, our equation is currently

$$\mathbf{M}\frac{d^2x}{dt^2} = \mathbf{K}\mathbf{x} + \mathbf{C}\frac{d\mathbf{x}}{dt} + \mathbf{F}(t)$$

where $\mathbf{M}$ is the mass matrix, $\mathbf{K}$ is the stiffness matrix, $\mathbf{C}$ is the damping matrix, $\mathbf{F}$ is the forcing matrix which only hits upon the first block, and $\mathbf{x}$ is a vector of the displacement of the blocks. The derivatives of the displacement can be placed as the finite differences between different measurements of position. Using the definition of the finite difference technique that we defined earlier, our equation becomes

$$\mathbf{M}\left(\frac{\mathbf{x}\langle t\rangle - 2\mathbf{x}\langle t - \Delta t\rangle + \mathbf{x}\langle t - 2\Delta t\rangle}{\Delta t^2}\right) = \mathbf{K}\mathbf{x}\langle t - \Delta t\rangle + \mathbf{C}\left(\frac{\mathbf{x}\langle t - \Delta t\rangle - \mathbf{x}\langle t - 2\Delta t\rangle}{\Delta t}\right) + \mathbf{F}\langle t\rangle .$$

Through algebra, we can rearrange the equation so that we have $\mathbf{x}\langle t\rangle$ as a function of everything else:

$$\mathbf{x}\langle t\rangle = \left(\Delta t^2\mathbf{M}^{-1}\mathbf{K} + \Delta t\mathbf{M}^{-1}\mathbf{C} + 2\right)\mathbf{x}\langle t - \Delta t\rangle - \left(1 + \Delta t\mathbf{M}^{-1}\mathbf{C}\right)\mathbf{x}\langle t - 2\Delta t\rangle + \Delta t^2\mathbf{M}^{-1}\mathbf{F}\langle t\rangle .$$

This is exactly where we want to be. Now, the future locations of the blocks, $\mathbf{x}_i$, can be calculated based upon their previous locations and the dynamics of the system. We can have an arbitrary forcing function, $\mathbf{F}\langle t\rangle$, and we can still find the solution.

Completing our example, we can substitute for our mass, stiffness, and damping matrices into our finite difference equation. *The result is a mess!* But, that is okay. Computers are great at keeping track of the mess. The finite difference programs are principally ways of bookkeeping all of the variables. There's also a bit of information about how to invert matrices and ways to efficiently update the grid[8] size, but mostly it is all about bookkeeping.

Since this is bookkeeping, we're going to skip the algebra steps. If you're bored, then you can substitute the matrices. We're only going to give the answer for the movement of the third block, $x_3$.

---

[8] This is computer talk for the size of the blocks into which we break up the system.

$$x_3\langle t\rangle = \frac{1}{m_3}\left(k_2\Delta t^2 + c_2\Delta t\right)\ x_2\langle t-\Delta t\rangle \ +\ \left(2+\frac{1}{m_3}\left(-k_2\Delta t^2 - c_2\Delta t\right)\right)\ x_3\langle t-\Delta t\rangle$$

$$+\ \frac{c_2}{m_3}\,x_2\langle t-2\Delta t\rangle\ +\ \left(1-\frac{c_2}{m_3}\right)\ x_3\langle t-2\Delta t\rangle$$

The position of the third block, $x_3$, is a function of the position of the second block, $x_2$. Thus, we can't directly find how the end of our pipe moves without determining how the pieces in the middle are moving. Also note that the current position of the block is a function of the previous positions of the blocks.

For simple situations, finite difference modeling can be performed using a spreadsheet such as Microsoft® Excel. However, you will probably want to use some of the prewritten software packages if the system is very complicated. At a minimum, you will use a command line program, such as Fortran, C, or Matlab®.

Assume that we want to know the acceleration at the right side of the pipe if we whack the left side of the pipe with a hammer. Let's examine how our system behaves with different number of elements in the pipe. Let's make the length of our two steel sections $L_s = 1$ meter, the central rubber section is $L_r = 1$ meter, and the cross-sectional area is $A = 0.01$ meter.

| STEEL | | |
|---|---|---|
| Modulus of Elasticity | $E_s$ | 200 GPa |
| Density | $\rho_s$ | 7900 kg/m3 |
| Damping | $\xi_s$ | 0.1% |
| **RUBBER** | | |
| Modulus of Elasticity | $E_r$ | 1 GPa |
| Density | $\rho_r$ | 900 kg/m3 |
| Damping | $\xi_r$ | 20% |

From engineering classes, we know that the stiffness of each element, $k$, is related to the modulus $E$ by

$$k = \frac{EA}{\Delta L}$$

where $\Delta L$ is the length of each block in the finite element and A is the cross-sectional area of the block. The mass of each element, $m$, is related to the density by

$$m = \rho A \Delta L .$$

There are many models for damping, and most of them are not very good. The gradual loss of energy due to friction and other forms of damping is difficult to accurately describe. We going to take a simple expression for damping, which is

$$c = 2\zeta \sqrt{\frac{E}{\rho}}.$$

Now we can actually look at the answer. Remember, we're trying to find the acceleration at the right side after we whack our beam on the left side. After 90 lines of very loosely written Matlab code, we can plot the behavior in acceleration in Figure 10-3. With our three-element model, you can see the oscillations in the acceleration due to the hammer strike. If you increase the number of elements, say to fifteen, then our model will more closely approximate the oscillations of a real structure. Increasing the number of elements in our model changes the answer a little bit, but we really don't need a lot of elements to get a good answer. This is very good, because the time that it takes to solve these types of problems increases with the square of the number of elements. And it's worse than that, because if we use smaller elements, then numerical stability requires us to use smaller time steps, which further increases the solution time.

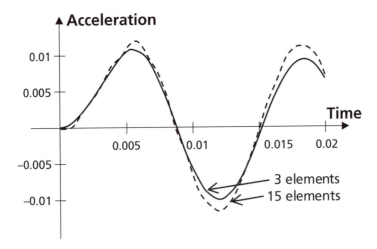

**Figure 10-3. Acceleration of three section system after 90 computational iterations.**

# Example: Experimental Data and Theoretical Equations

We've already shown an example using the theoretical equation for the composition of alloy crystals in the early stages of growth. The equation involved both growth parameters which the scientist may vary, and it included parameters which Mother Nature holds to herself. The equation was in the form of an exponential. If a dedicated scientist wanted to match data to theory, the scientist could get out the log tables and in a few hours generate a reasonable curve to plot against the data.

The equation[9] for the end of growth is not so benign. It is

$$C(x) = C_0 \left[ 1 + 3\left(\frac{1-k}{1+k}\right) e^{-2(R/D)x} + 5\left(\frac{(1-k)(2-k)}{(1+k)(2+k)}\right) e^{-6(R/D)x} \right.$$
$$\left. + \cdots + (2n+1)\left(\frac{(1-k)(2-k)\cdots(n-k)}{(1+k)(2+k)\cdots(n+k)}\right) e^{-n(n+1)(R/D)x} + \cdots \right.$$

where

D and k are materials properties

R is the growth rate

and

x, the crystal length, is measured from the last-to-grow end of the crystal.

As you can see, there is nothing here that can't be solved with sweat and log tables, but it would not only be laborious and tedious to an extreme, it would also be fraught with potential errors. The computer makes such theoretical analysis useful.

---

[9] See, "A Mathematical Analysis of Solute Redistribution During Solidification," V. G. Smith, W. A. Tiller. and J. W. Rutter, *Canadian Journal of Physics*, vol. 33, pages 723-745, 1955.

CHAPTER **11**

# *Chaos*

*So inscrutable is the arrangement of causes and consequences in this world that a two-penny duty on tea, unjustly imposed in a sequestered part of it, changes the condition of all its inhabitants.* — **Thomas Jefferson, Autobiography**

Don't let this title mislead you. Although *Random House Webster's Dictionary* gives the first definition of chaos as "a state of utter confusion and disorder" and the second definition as "any confused, disorderly mass," it goes on to mention our magic words for chaos as "deterministic behavior." This has been the function of engineering and engineering math all along: to determine the behavior of physical systems. In this chapter we'll combine physics, philosophy, and mathematics.

"Wait a minute," you might complain. "Physics and physical principles upon which engineering is based are deterministic." If we know the properties of our system, and if we know physics, we can determine how something will behave. This type of thinking started back in the sixteenth century when Copernicus defined our solar system and got Galileo thinking about the scientific method. This led Newton to formulate three simple laws that not only describe the motion of Galileo's cannon balls falling from the Leaning Tower of Pisa but also describe the motion of our solar system. "Science," you might continue, "is not a state of confusion. It's order."

Well, maybe and maybe not. It depends on how you look at some things. Let's consider a few experiments.

*Experiment 1. Falling Cannon Balls*

If you drop a cannon ball from a tower[1], you can reasonably calculate the transit time. If you repeat the experiment, assuming that you're not arrested for dropping cannon balls, you can expect the results to repeat to a high degree of accuracy and be very close to your calculated value.

You, however, cannot determine beforehand whether a strong wind might cause up-drafts at the base of the tower and perturb your measurements. If such wind does occur, you'll still repeat your initial measurements with only a small error.

---

[1] If you must follow the scientific principle and perform the experiment, don't go to Pisa. There are too many people in the square. Find an abandoned fire tower somewhere in the woods.

*Experiment 2. Voltage–Current Measurements*

Take a 10-volt battery and place a 2000-ohm resistor across the terminals. The ammeter in your circuit reads[2] 0.005 amps. This is the value you predicted. However, if you get an ammeter of higher accuracy, you may see the current varying as the undetermined humidity in your lab changes or as the battery loses its charge.

*Experiment 3. Boats in the Stream*

Take a few toy boats and put them into your backyard stream. Although all of the boats are the same, you place them in the water the same way, and although the stream does not seem to vary in its overall flow, the boat trajectories are seldom the same. These experiments do not repeat.

What happened? Do cannon balls and simple electrical circuits follow physical laws, but floating boats do not? Are only experiments 1 and 2 deterministic? Even though they had errors from causes that we could not *a priori*[3] predict, we can *a posteriori* explain what happened.

Do not worry. The laws of fluid mechanics are well in hand. It's just their application that requires great care. Your backyard stream and the boats represent a dynamic system with extreme sensitivity to initial conditions. We call such systems chaotic. If you measure the position and speed of each toy boat a short distance from release, you'll get reasonable[4] repetition of results. The further downstream you go, the more chaotic the results.

Long-term mathematical predictions of the behavior of chaotic systems are no more accurate than random chance. Short-term predictions, however, can be accurate. Another well-known example of this extreme sensitivity to initial conditions is the weather. Small changes in weather patterns can result in large changes later on. We're sure you've noticed how the weather folks (who use copious fluid dynamics equations in their calculations) can reasonably predict rain tomorrow, but they can't predict if it will be raining on a given day next year.

---

[2] For the nonelectrical engineers in the audience, we are using Ohm's Law which states that current (in amperes or amps) is equal to voltage (in volts) divided by resistance (in ohms).

[3] Since we said this chapter delves into philosophy, we decided to add some Latin. *A priori* means "conceived beforehand." We bet that you can figure out the meaning of *a posteriori*.

[4] We hope you'll forgive our lack of quantification in these examples. What are "a short time" and "reasonable results" depend on more parameters than we need to get into and also depend on your desired accuracy of results. You'll soon see mathematically quantified results varying far more than you may have predicted, and these qualifications will make more sense then.

Tossing coins and rolling dice, as discussed in the chapter on probability and statistics, are also examples of chaotic behavior. Even though the tossing or rolling is completely deterministic, you must know the exact orientation and momentum of the coins or dice and the exact compliance of the landing surfaces to know whether you'll win or lose. However, those variables are not known exactly, thus your results are unknown.

"What about the short-term behavior?" you ask. "Even chaotic systems are predictable on a short-term basis."

Yes, you are right, and with the coins and dice, short-term does not refer to the first few results of the tosses or rolls but to the first few revolutions of the coin or die which, if you are careful, can be reasonably predicted.

For a mathematical example of deterministic chaos, let's take the simple recursive[5] equation[6]

$$x_{i+1} = 2x_i^2 - 1$$

where we are iterating on $x_i$. That is, we take the value just calculated for $x_i$ and put it into the equation to make the next calculation and determine a value for $x_{i+1}$. For example, if our first value of $x$ were 0.75, then our new value for $x$ would be

$$x_2 = 2x_i^2 - 1$$
$$= 2 * (0.75)^2 - 1$$
$$= 0.125$$

and $x_3$ would be calculated from $x_2$, and so on.

That seems easy enough. Now, program your computer[7] to calculate another 23 values of $x_i$ so that you have a total of 25 values of $x$.

These repeated calculations may seem tedious to you, but repetition and iteration of calculation is the stuff of chaos, and computers don't get bored. You don't see the nature of the extreme sensitivity to initial conditions until you've gone through a few steps.

After you list those 25 values of $x$, repeat the process after changing your initial condition. That is, change your value of $x_1$ by 1%. You would expect a small change in your calculations, and sure enough, $x_2$ is in error by 18%. If that seems like a large error for a 1% change in initial conditions, look at $x_8$. It's off by 500%. OK, a 500% error after just seven iterations is not acceptable. That's analogous to the meteorolo-

---

[5] In a recursive equation, each new term is determined by the application of an equation that *uses preceding results as the variable*.

[6] This is close to a Mandelbrot or Julia set. The general form for these sets is $x_{i+1} = x_n^2 + C$, ($C$=constant and $x$ is complex). The difference in these two sets is determined by the constant and the initial conditions.

[7] A spreadsheet works just fine for this problem.

gists being off by 500% after just seven steps in calculating weather patterns. So you refine your measurements so that your initial measurement is accurate to 0.1%, and you repeat your calculations. Everything starts out looking much better. Well, it looks better for the first few iterations. As the error of the initial condition decreases by a factor of ten, the error in the results also drops by approximately the same factor for the first seven iterations. By step eleven, however, the error is greater with the more precise initial condition. Well, it seems that we're getting better. We determined more precise initial conditions, and we got better results (at least for a while). Can we do better? Yes, but only for a little while.

We want to use super-precise measuring instruments. We buy the best equipment available, and build what we can't buy. We measure our initial condition to one part in a million. That's a precision of one second in eleven and a half days. That precision is possible with time measurement but not with much else, but we do it. And look at the chart: the first fifteen steps are accurate to within 5%. Then it blows up. We won't waste the paper here, but in 100 steps the average error was 110% for the 1% measurement, 111% for the 0.01% measurement, and 171% for the one part in a million initial measurement.

### Table 11-1. Demonstration on initial condition sensitivity

| Step | Value of $X_i$ | % error for 1% change in initial x | % error for 0.1% change in initial x | % error for 0.0001% change in initial x |
|---|---|---|---|---|
| 1 | 0.750 | 1.00% | 0.10% | 0.0001% |
| 2 | 0.125 | 18.09% | 1.80% | 0.00% |
| 3 | −0.969 | −1.27% | −0.12% | −0.00% |
| 4 | 0.877 | −5.41% | −0.50% | −0.00% |
| 5 | 0.538 | −30.11% | −2.86% | −0.00% |
| 6 | −0.421 | 70.37% | 7.76% | 0.01% |
| 7 | −0.646 | −104.42% | −8.84% | −0.01% |
| 8 | −0.166 | 500.62% | 84.80% | 0.09% |
| 9 | −0.945 | −205.16% | −14.13% | −0.01% |
| 10 | 0.785 | 24.07% | −59.70% | −0.05% |
| 11 | 0.233 | 285.81% | −443.80% | −0.49% |
| 12 | −0.892 | −168.55% | −131.35% | 0.12% |
| 13 | 0.590 | −142.80% | −242.90% | 0.63% |
| 14 | −0.303 | 188.09% | −239.94% | −2.93% |
| 15 | −0.817 | −163.89% | −21.52% | 1.30% |
| 16 | 0.334 | −236.46% | −153.44% | 10.42% |
| 17 | −0.777 | −24.73% | 20.49% | −6.29% |
| 18 | 0.208 | −252.04% | 262.71% | −70.83% |
| 19 | −0.914 | −12.39% | −114.81% | 8.64% |
| 20 | 0.670 | −57.95% | −243.85% | 44.95% |
| 21 | −0.103 | 716.74% | −931.11% | −958.77% |
| 22 | −0.979 | −142.48% | −147.62% | −157.75% |
| 23 | 0.916 | −171.42% | −161.75% | −139.40% |
| 24 | 0.678 | −121.24% | −153.12% | −209.04% |
| 25 | −0.080 | 1093.71% | 822.21% | −216.46% |

As you can see from the table of results, the more precise initial measurement sometimes gives the greater error. Are we dealing with an erratic computer? Does math no longer work? Is this simple equation not deterministic? The answer is an emphatic "No!" to all three questions. You have just seen an example of ***mathematical chaos***. If you repeat the calculations *with the same initial conditions*, you'll get the same answers. The equation is quite deterministic. It is just extremely sensitive to the initial[8] conditions.

The idea that such small perturbations in initial conditions can have such large effects later on is sometimes called the *Butterfly Effect*. Some folks say, and not totally in jest, that whether or not a butterfly flaps its wings in a certain part of the world can make the difference in whether or not a storm arises later in another part of the world. No, we don't really believe the literal Butterfly Effect, but you can see the reason that weather forecasts can be accurate only in the short-term, and that long-term forecasts, even using the best computer methods available, are no better than the *Farmer's Almanac*.

Does chaos only apply to the weather and simple equations as shown? No, it applies to any nonlinear, coupled system. Our simple equation

$$x_{i+1} = 2x_i^2 - 1$$

applies to no physical system that we know of. It is just a demonstration tool that is nonlinear (the $x_i^2$ term makes it nonlinear), and is coupled with itself (as the previous value is used to calculate the current value).

Complex fluid flow systems (which include weather calculations) include an equation for the conservation of momentum, another one for the conservation of energy, and a third for the conservation of material in the flow. These equations are nonlinear except in their simplest compilations, and they are often coupled since each equation contains terms determined by the other equations. These equations[9] are studied in advanced fluid dynamics courses, and we mentioned one of them in our differential equations chapter. Just remember that you first heard it here.

Another example of sensitive initial conditions and wide ranging results is the predator-prey model. Here we are dealing with a predator, such as a bird, and its only prey, a special butterfly. The birds must have butterflies (BF in the equation) to survive, but too many birds deplete the butterfly supply and birds starve—it's a tough world out there. We formulate the model[10] as follows:

---

[8] It's not just the first stated initial conditions that drive chaos. Should you truncate one of your values anywhere down the line, subsequent values will go chaotic the same as if the first (initial) value were in error. With iterative equations, any value, anywhere down the line, is the initial value for all following calculations.

[9] Navier-Stokes equations.

[10] The biologists in the audience may scream at the simplicity of this model – sorry.

BF this year = BF last year · (1+ BF growth rate – Bird kill rate · BF last year)

Birds this year = Birds last year · (1+ Bird growth rate · BF last year – Bird death rate)

where we abbreviate butterflies as BF and assume that butterflies die only as they provide lunch for hungry birds.

We could graph a few examples for you, but this is an easy one for you to do yourself with a spreadsheet.

Start with these values

Birds for the first year = 50  (i.e., "Birds last year" in the first iteration)

Butterflies for the first year = 200 (i.e., "Butterflies last year" in the first iteration)

Butterfly growth rate = 0.05

Bird kill rate = 0.001

Bird growth rate = 0.0002

Bird death rate = 0.03,

Iterate a few hundred times, and graph the results of the number of birds and butterflies as a function of the iteration step. Change the input conditions by what you consider to be a small amount. Notice how the graph radically changes.

Now plot your graph a little differently. The number of birds and the number of butterflies for each period make a set of numbers $(x,y)$; either birds or butterflies can be $x$. Plot these values as $x$ vs. $y$. Notice that you get spirals of the growing and diminishing populations. But regardless how the populations vs. time period curve changes as you change input parameters, the bird population vs. butterfly population graph shows only small changes. You are now playing with a ***fractal pattern***.

A fractal has two primary properties.

1.  It is a geometric object that is similar to itself on all scales.

2.  It has a fractional dimension. The fact that a fractal has the same shape at different dimensions[11] is seen in the birds vs. butterflies graphs. Regardless how many times you iterate on their population variations, the curve looks basically the same. We will go to another example to talk about the fractional dimension.

---

[11] These plots are often called phase diagrams, which comes from mechanical systems where position of the subject (the function) is plotted against the velocity of the subject (the first derivative of position). You saw a phase diagram in differential equations.

You have all seen pictures of snowflakes. The single crystal of a snowflake is formed by sides growing upon sides[12] such that all sides look similar, but each one may seem a bit different. We won't use a real snowflake, but we'll use a model, the Koch snowflake,[13] to explain the concept of fractional dimension. You, too, can draw a Koch snowflake. First draw a line one unit long as shown in Figure 11-1. Then replace that line with four lines, also shown, where each of the four lines are one-third the length of the original line. The total length of the new line is four-thirds longer than the original line, but the horizontal dimension has not changed. Repeat the process on each of the four new segments, and continue on and on.

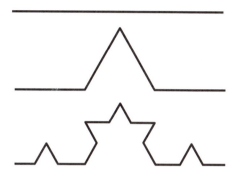

**Figure 11-1. A Koch snowflake.**

Had you started with three segments arranged in a triangle instead of the single line, your Koch snowflake would take on the shapes shown in the Figure 11-2. Note that the basic snowflake looks the same regardless of how you zoom in on it. That is, it continues to look like the original shape. As an aside, notice how the length of the perimeter grows without bound as the enclosed area changes very little.

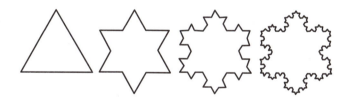

**Figure 11-2. Koch snowflake starting with a triangle.**

Before we determine the dimension of the Koch snowflake, let's define dimension. We know that a line is one-dimensional, and a square is two-dimensional. Let's further quantify that concept.

---

[12] Materials scientists call this dendritic growth.
[13] First drawn by Helge von Koch in 1904.

If we take a line and increase its length by a factor $N$, the ratio, $r$, of each new segment to the original length will be

$r = N$

If we increase the area of the square by the factor $N$, the length of each new side will increase by

$r = N^{1/2}$ .

That is, if you increase the area of a square by four times its original area, each new side will be twice the length of the sides of the original square.

Nothing new so far. Let's write *dimension* in a different way. The ratio of new side length to the original side length can be written as

$r = N^{1/D}$

or

$$D = \frac{\log N}{\log r}$$

where $D$ is the dimension of your object. (Log to any base works.)

Now, let's apply our new formulation of "dimension" to the snowflake. Notice that each time you expanded the snowflake you made a length of three units into a length of four units. Hence, the dimension of the Koch snowflake is

$$D = \frac{\log 4}{\log 3}$$

$D = 1.26...$

The dimension of this design is a fraction, which makes the design a fractal[14].

We will work one more set of recursive equations that generate an unexpected fractal pattern. We'll play with Sierpinski's triangle. Sierpinski's triangle generates an endless number of x-y values which we'll plot as $x$ vs. $y$. But Sierpinski's triangle, though recursive, uses three different sets of equations where the choice of equation is determined by random numbers. Sounds confusing? It isn't. Just follow the rules:

---

[14] Fractals have also been used to examine surface erosion (Furhrmann, John, "The Fractal Nature of Erosion: Mathematics, Chaos, and the Real World", *Erosion Control*, June 2002). Researchers have shown that the cyclical erosion and deposition process which occurs on beaches through tide cycles or in stream and rivers responding to seasonal variations in flow can be assessed with cyclical fractal growth techniques (Shapir, Y. S. Raychaudhuru, D. G. Foster, and J. Jorns. "Scaling Behavior of Cyclical Surface Growth" *Physical Review Letters*, 84(14): 3029-32, 2000). While much of this type of analysis is in its infancy, it holds promise for better understanding the physical world in the future.

1. Choose a random number.

2. If the random number is less than 1/3, use the following equations to determine $x$ and $y$.

   $x_{n+1} = 0.5(x_n + 1)$ and

   $y_{n+1} = 0.5 y_n$

3. If the random number is between $\frac{1}{3}$ and $\frac{2}{3}$, use these equations to determine $x$ and $y$.

   $x_{n+1} = 0.5 x_n$ and

   $y_{n+1} = 0.5 y_n$

4. If the random number is greater than $\frac{2}{3}$, then

   $x_{n+1} = 0.5(x_n + 0.5)$ and

   $y_{n+1} = 0.5(y_n + 1)$.

Program your computer to run out a few hundred values of $x_i, y_i$ and plot $x_i$ vs. $y_i$. Plot only the computed points. Do not plot any connecting lines. You'll see fascinating, nested sets of triangles developing. Figure 11-3 is a Sierpinski's triangle with 4000 data pairs generated with a spreadsheet. Do it yourself. It's fun!

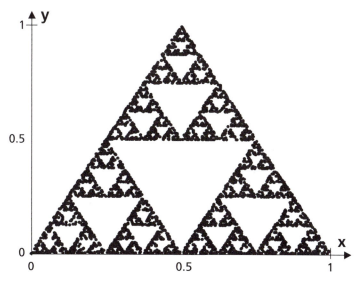

**Figure 11-3. Sierpinski's triangle with 4000 data pairs.**

Does Sierpinski's triangle look like modern art? Maybe. Jackson Pollock's "paint-drip" modern art paintings have been scientifically examined[15] and found to have fractal dimensions. The majority of his paintings have a fractal dimension of approximately 1.6.

Let us leave this chapter with a thought and a hope. We have briefly discussed the scientific philosophy of determinism. Now let's dabble into the philosophy of life. If small perturbations and uncertainties in starting points can have tremendous effects later on physical systems, does it seem reasonable that the apparently insignificant events in a person's life can produce important consequences much later? How did you meet the significant other in your life? Did you have a plan? Did your parents hire a matchmaker? Not likely. You probably met only because the two of you happened to coincidentally be at the same place at the same time[16]. Had you taken a different path that day, think how different your life would be.

Consider another possible example of the chaos of life. Assume that one morning on your way out the door, you were asked to drop the trash into the receptacle by the curve. This simple act (initial condition) took 30 seconds. Because of that brief lag in the start of your day, you missed the green light at the corner and had to stop. While waiting for the light to change, a preoccupied stockbroker rammed into the back of your car. Contrite at causing the accident, the stockbroker told you about the cause of the mind-slip: a tip on a great stock[17] to purchase. You buy the stock, become a multimillionaire, and now grow orchids on your ranch in Hawaii. All that started because of a 30-second trip to the trash can. And if you're not a millionaire, perhaps it's because of a small change in another initial condition—you did not take out the trash one day when you should have.

Please remember: Newton's laws are still valid. There is no randomness in the equations of motion. They are completely deterministic. Chaos—that appearance of the vague outcome—comes only from the lack of absolute accuracy in the knowledge of the initial conditions.

We only want to introduce the subject of chaos to let you know that there's a lot more out there[18] in the fascinating world of engineering mathematics. We leave this chapter with the hope that all of the chaos of your life dissolves into beautiful fractal patterns.

---

[15] R. P. Taylor, *Scientific American*, December, 2002, pp. 116–121.

[16] One of the authors met his future wife because each one went to the *wrong* chemistry class, but that's another story.

[17] Some folks argue that the stock market is truly chaotic. We'll leave that for the business majors to ponder.

[18] For further reading see: Gleick; *Nonlinear Dynamics and Chaos*, Steve Strogatz, 1994, Westover, Perseus Books, Cambridge, MA; and *Chaotic and Fractal Dynamics*, Francis Moon, 1992, John Wiley & Sons, NY, NY.

# Some Useful Mathematical Tables

**Table A-1. Most frequently used derivatives**
**(where $c$ = a constant and $u$ and $v$ are functions of $x$).**

$$\frac{dc}{dx} = 0$$

$$\frac{dx}{dx} = 0$$

$$\frac{d}{dx}(u + v) = \frac{du}{dx} + \frac{dv}{dx}$$

$$\frac{d}{dx}(cu) = c\frac{du}{dx}$$

$$\frac{dy}{dx} = \frac{dy}{du} \cdot \frac{du}{dx}$$

$$\frac{d}{dx}\log_a u = \frac{1}{u}\log_a e\frac{du}{dx}$$

$$\frac{d}{dx}\ln u = \frac{1}{u}\frac{du}{dx}$$

$$\frac{d}{dx}(u^n) = nu^{n-1}\frac{du}{dx}$$

$$\frac{d}{dx}(uv) = v\frac{du}{dx} + u\frac{dv}{dx}$$

$$\frac{d}{dx}\left(\frac{u}{v}\right) = \frac{v\frac{du}{dx} + u\frac{dv}{dx}}{v^2}$$

$$\frac{d}{dx}\sin u = \cos u\frac{du}{dx}$$

$$\frac{d}{dx}\cos u = -\sin u\frac{du}{dx}$$

$$\frac{d}{dx}\tan u = \sec^2 u\frac{du}{dx}$$

$$\frac{d}{dx}\cot u = -\csc^2 u\frac{du}{dx}$$

$$\frac{d}{dx}\sec u = \sec u \tan u\frac{du}{dx}$$

$$\frac{d}{dx}\csc u = -\csc u \cot u\frac{du}{dx}$$

$$\frac{d}{dx}a^u = a^u\ln a\frac{du}{dx}$$

$$\frac{d}{dx}e^u = e^u\frac{du}{dx}$$

$$\frac{d}{dx}u^v = vu^{v-1}\frac{du}{dx} + u^v\ln u\frac{dv}{dx}$$

## Table A-2. Most frequently used integrals
### (where *c* and *a* are constants and *u* and *v* are functions of *x*).

$$\int du = u + c$$

$$\int u^{-1} du = \int \frac{du}{u} = \ln|u| + c$$

$$\int \sec u \tan u\, du = \sec u + c$$

$$\int (du + dv) = \int du + \int dv$$
$$= u + v + c$$

$$\int \sin u\, du = -\cos u + c$$

$$\int \csc u \cot u\, du = -\csc u + c$$

$$\int \cos u\, du = \sin u + c$$

$$\int a^u du = \frac{a^u}{\ln a} + c$$

$$\int adu = a\int du = au + c$$

$$\int \sec^2 u\, du = \tan u + c$$

$$\int e^u du = e^u + c$$

$$\int u^n du = \frac{u^{n+1}}{n+1} + c$$

$$\int \csc^2 u\, du = -\cot u + c$$

## Table A-3. Useful Conversion Factors.

| | | |
|---|---|---|
| acre | = | $0.0016 \text{ mile}^2 = 4.3560 \times 10^4 \text{ ft}^2$ |
| centimeter | = | $0.0328 \text{ ft} = 0.3937 \text{ in}$ |
| cubic foot | = | $7.4805 \text{ gal} = 0.0283 \text{ m}^3$ |
| degree | = | $0.0175 \text{ rad}$ |
| foot | = | $0.3048 \text{ m}$ |
| gallon | = | $0.1337 \text{ ft}^3 = 3.7853 \text{ liters}$ |
| inch | = | $2.5400 \text{ cm}$ |
| kilogram | = | $2.2046 \text{ lb (avdp)}$ |
| kilometer | = | $0.6214 \text{ mile}$ |
| liter | = | $0.2642 \text{ gal} = 0.03532 \text{ ft}^3$ |
| meter | = | $1.9036 \text{ yd} = 3.2808 \text{ ft} = 39.3701 \text{ in}$ |
| mile | = | $1.6093 \text{ km}$ |
| ounce | = | $1.8047 \text{ in}^3$ |
| pound (avdp) | = | $0.4536 \text{ kg} = 453.5924 \text{ gm}$ |
| pound force | = | $4.4482 \text{ newtons}$ |
| quart | = | $0.9463 \text{ liter}$ |
| radian | = | $57.2958 \text{ degrees}$ |
| yard | = | $0.9144 \text{ m}$ |

### Table A-4. Some Numerical Constants.

| | | |
|---|---|---|
| $e$ | = | 2.71828 18285 |
| $1/e$ | = | 0.36787 94412 |
| $e^2$ | = | 7.38905 60989 |
| $e^{1/2}$ | = | 1.64872 12707 |
| $e^{-1/2}$ | = | 0.60653 06597 |
| p | = | 3.14159 26536 |
| $1/p$ | = | 0.31830 98862 |
| $p^2$ | = | 9.86960 44011 |
| $p^{1/2}$ | = | 1.77245 3850 |
| $\log_e p$ | = | 1.14472 98858 |
| $\log_{10} e$ | = | 0.43429 44819 |
| $2^{1/2}$ | = | 1.41421 35624 |
| $3^{1/2}$ | = | 1.73205 08076 |
| $10^{1/2}$ | = | 3.16227 76602 |
| $(2p)^{1/2}$ | = | 2.50662 82746 |
| $(p/2)^{1/2}$ | = | 1.25331 41373 |

## Table A-5. Greek alphabet.

| Case | | English |
|:---:|:---:|:---:|
| **Upper** | **Lower** | **Name** |
| A | α | alpha |
| B | β | beta |
| Γ | γ | gamma |
| Δ | δ | delta |
| E | ε | epsilon |
| Z | ζ | zeta |
| H | η | eta |
| Θ | θ | theta |
| I | ι | iota |
| K | κ | kappa |
| Λ | λ | lambda |
| M | μ | mu |
| N | ν | nu |
| Ξ | ξ | xi |
| O | o | omicron |
| Π | π | pi |
| P | ρ | rho |
| Σ | ς | sigma |
| T | τ | tau |
| Y | υ | upsilon |
| Φ | φ | phi |
| X | χ | chi |
| Ψ | ψ | psi |
| Ω | ω | omega |

# About the Authors

Archie Fripp has been fascinated by mathematics ever since he bought two ten-cent ice cream cones with the quarter he earned mowing a neighbor's lawn. He was excited to have a nickel left to spend. Looking for more technically challenging jobs, he joined the Air Force in his mid-teens and spent four years working on radar sets at various bases and installing communication cables on NASA and Air Force gantries at Cape Canaveral. Following Air Force duty, he sailed the two largest oceans on a NASA ship tracking the Mercury space flights.

After those adventures, he decided to upgrade his credentials once more. With a B.S. from the University of South Carolina in electrical engineering and an M.S. and Ph.D. in materials science from the University of Virginia, he found himself mowing his own grass on the weekends while he worked at the NASA Langley Research Center in Hampton, Virginia.

At Langley, he worked in basic research on semiconductor devices and served as a consultant for space flight projects, including the Viking Lander that now sits on Mars and the Hubble Large Space Telescope that now searches the heavens. Through a world-wide competition, Archie became the Principal Investigator on a series of microgravity science experiments growing semiconductor crystals on the Space Shuttle.

Archie is now enjoying retirement writing a math book, trying to capture the beauty of this world through photography, and mowing the lawn.

Growing up with Archie as a father, Jon and Michael were always interested in math. As a family, they entertained each other with math problems on their long car trips. Jon was particularly interested in applying math to his high school summer jobs, building docks and houses. Michael was more interested in math for its own sake. In high school, Michael's definition of a "fun weekend" was one spent at a math competition.

Jon, now a registered professional engineer, went to Virginia Tech and got B.S. and M.S. degrees in civil engineering. Unable to resist building things as well as stomping in the mud, Jon started work as a hydraulic engineer for the Army Corps of Engineers. Now with the USDA Natural Resources Conservation Service in Fort

Worth, Texas, Jon applies his mathematical skills and intuition towards hydraulic and environmental projects across the United States. Jon and his family still enjoy the building projects on their home and ranch.

Michael followed his big brother to Virginia Tech and got a B.S. in engineering science and mechanics. Moving northward, he got an M.S. and a Ph.D. in aeronautical and astronautical engineering from Massachusetts Institute of Technology. In one math class, he found more than just a love for math; he also found his wife. Upon graduation, Michael opted to become a rock scientist rather than a rocket scientist and went to work for Halliburton Energy Services in Carrollton, Texas. Michael continues to enjoy math for its own sake, developing mathematical models for the problems encountered in the petroleum industry. Michael and his family still entertain each other with math problems on long car trips.

The Fripp family likes math. They use math both professionally and in their daily lives. They hope you enjoy the application and the beauty of mathematics as you use this book.

# Index

## A

a posteriori, 326
a priori, 326
abscissa, 24
absolute value, 9-10, 144
acceleration, 129-130, 168-171
adding trigonometric functions, 94-98
addition, 2
algebra, 37-58
angle, 83
    acute, 84
    complementary, 83, 84
    measurement, 85-87
    obtuse, 84
    supplementary, 84, 88
arctangent, 102
associative law
    addition, 4
    multiplication, 5
axes, 23

## B

base 2, 15-16
base, 10, 15
bell curve, 202
Bernoulli, 198
binary number system, 16
binomial distribution, 196-199
boundary value problems, 250-252
butterfly effect, 329

## C

calculus, 125-176
    applications of, 168-176
    differential, 126-137
    integral, 148-151
cancellation rule, 8
Cartesian coordinate system, 23
center of mass, 203
Chain Rule, 137
chaos, 325-334
characteristic equation, 235
chi-square test, 219-220
coefficient of friction, 118
combinations, 189
commutative law
complementary angle, 83, 88
complementary function, 234
completing the square, 48
complex conjugate, 50
complex numbers, 7-8, 50
compound function, 137-143
computer mathematics, 311-323
confidence interval, 220-221
contingency table, 195
continuity equation, 287-288
continuous function, 144
convergence, 145
conversion factors, table, 337
coordinates, 24, 26
    transformation of, 293
correlation, 213-217

cosine, 90

cost function, 316

Coulomb's Law, 302

cross product, 265, 272-275

curl, 284, 290-292

current loop, force on, 302-305

curve fitting, 217-219

cylindrical coordinate, 172

**D**

damping, 239-243

data analysis, computerized, 311-312

decimal, 4

definite integral, 153

degree of a polynomial, 40

dependent equations, 62

dependent variable, 20-22

derivative of a function, 130-137

    compound functions, 137-143

    polynomials, 130-133

    table of common, 335

determinant, 72, 276

diagonal matrix, 77

differential calculus, 125, 126-137

differential equations, 227-263

    general solution, 232

    multiple roots of, 243

    particular solution, 234-235

    series solutions to, 252-253

differentiation, computer, 314

dimensionless variable, 104

direct integration, 248-250

discretized vs. digitized, 313

discriminant, 49

dissipation, 111-112

distance between two points, 24

distance equation, 92, 93, 269

distribution function, 183-184

distributions, 196-199

distributive law, 5

Divergence Theorem, 284

divergence, 284

division, 3

dot product, 265, 270-272

**E**

electrical circuit analysis, 79-80

electromagnetic theory, vectors in, 297-301

equality, 2

equations with more than two variables, 74-76

equipotential lines 282

Euler column, 250

Euler-de Moivre formula, 233

event, 178

Excel, Microsoft, 322

experiment, 178

exponents, 11

**F**

factorial, 186

factoring, 40, 51

finite difference technique, 312, 320

finite element techniques, 318

first derivative, 141

first moment, 204-206

first-degree polynomials, 43

flux lines, 288

Fourier analysis, 159-162

Fourier series, 146

fractal pattern, 330

fractions, 4

free-body diagram, 114, 319

frequency, 105

    function, 183

    shift, 110

friction, 118-120

functions, 19-35, 125

    graphing, 20, 23-24

    periodic, 103

    trigonometric, 90-91

Fundamental Theorem of Calculus, 151-156

**G**

Gauss's Law, 301
Gaussian distribution, 207-210, 224-226
general solution to differential equation, 232
gradient descent optimization, 317
gradient, 282
graphing
    two linear equations, 59-61
    functions, 20, 23-24
gravity, 227
Greek alphabet, 339

**H**

homogeneous equations, 74, 234
hypotenuse, 90

**I**

identities, trigonometric, 98
imaginary numbers, 7, 50
inconsistent system, 63
indefinite integral, 153
independent functions, 236
independent variable, 20-22
inequality symbols, 8-9
inflection point, 142-143, 212
inner product, 270
integers, 2
integral calculus, 126, 148-151
integrals
    line, 288
    table of common, 336
integration, 153
    by parts, 154-156
interior angles, 99-101
interval of existence, 145
isosceles triangle, 99

**J**

$j$ in electrical engineering, 7, 50

**K**

Kronecker delta, 271
Koch snowflake, 331

**L**

L'Hopital's Rule, 147
Law of Sines, Cosines, 101-103, 115
least squares, method of, 217
limits, 144
line integrals, 288-290
linear differential equations, 234
    with constant coefficients, 235
linear differential operator, 234
linear equations and functions, 28, 44
log functions, 33-36
log plots, 35-36
log tables, 14
logarithms, 13-15
Lorentz Force, 301-302

**M**

Maclaurin's expansion, 146
Matlab, 321-322
matrix, 59-81, 319
maxima, 104, 140, 212
measurement, 211
method of least squares, 217
minima, 104, 140, 316
moment, 202
Monte Carlo simulation, 317
multiplication, 3

**N**

natural logarithm, 134
Navier-Stokes equations, 261-263
net circulation integral, 288
nonhomogeneous differential equation, 244-248
nonlinear differential equation, 234
nonlinear functions, 32
normalized data, 215

numbers, 2-4, 6-7
  bases, 15
  complex, 7-8
  imaginary, 7
  irrational, 6
  negative, 2
  rational, 6
numerical constants, table, 338

**O**
optics, 120
optimization techniques, 316-318
ordered sets, 26, 44
ordinate, 24
origin, 23
orthogonal curvilinear coordinates, 292-297
orthogonal unit vectors, 269, 281, 283
oscillating force on spring-mass-damper system, 244
outcome, 178

**P**
partial derivative, 254
Pascal's triangle, 199-202
periodic function, 103
periodicity, 103
permutation, 186
phase difference, 107-110
phase shift, 107
phase, 107, 109
plane, 278
plotting functions, 23-24
Poisson distribution, 209-213
poles, 51
polynomials, 39-40
  degree of, 40
  dividing, 41
  multiplying, 41
  second-degree, 46-54
pressure, 171
probability distribution, 204

probability, 177-195
projectile motion, 305-309
Pythagorean theorem, 25, 87, 102

**Q**
quadrangle, 99
quadrant, 84
quadratic equations, 46, 51
quadratic formula, 49

**R**
radian, 86, 88-89
reciprocal, 8
recursion formula, 199
reflection, 120-121
refraction, 120-121
regression analysis, 217-219
regression line, 218
right triangle, 25, 90
right-hand rule, 268
root mean square, 156, 167
roots, 12, 51

**S**
scalar matrix, 69
scalar product, 270
scalar, 266, 268
  solving for values, 269-270
second derivative, 141-143
second moment, 204
second-degree polynomials, 46-50
sequence, 145
series, 144
shotgun approach to finding minima, 317
Sierpinski's triangle, 332-333
signal noise, 316
sine, 90
  law of, 101
  wave, 104
slope, 26-28
smooth function, 144
Snell's Law, 122

solving for x, 39
solving systems of equations, 62-70
    graphing method, 59-66
    matrix method, 68-74
    substitution method, 66-68
spherical coordinates, 294
spring constant, 228
spring-mass-damper system, 319
square matrix, 69
square root, 50
standard deviation, 211
statics, 113-117
subtraction, 2
superposition, 232
supplementary angle, 84, 88

**T**
tangent, 90
Taylor's expansion, 146
term, 41
transformation of coordinates, 193
translation of coordinates, 31
triangle
    isosceles, 99
    right, 25, 90
triangular form, 69
trigonometric distance equation, 93
trigonometric identities, 98
trigonometry, 83-123
triple scalar product, 276
triple vector product, 277

**U**
units, 3, 63, 129
unit vector, 268-269

**V**
vector calculus, 265-309, 280-282
vectors, 113, 266
    magnitude, 267
    resultant, 267
    products, 270-275
velocity, 127-130
vibration, 318

**W**
wave, sine, 104
wavelength, 104
Wronskian, 237

**X**
x-axis, 24
x-ray diffraction, 108

**Y**
y-axis, 24
y-intercept, 6, 28-29

**Z**
zero, 2, 6
    of an equation, 51